Valuing the Built Environment

GIS and house price analysis

SCOTT ORFORD
School of Geographical Sciences, University of Bristol

Ashgate

Aldershot • Brookfield USA • Singapore • Sydney

Published by
Ashgate Publishing Ltd
Gower House
Croft Road
Aldershot
Hants GU11 3HR
England

Ashgate Publishing Company
Old Post Road
Brookfield
Vermont 05036
USA

Ashgate website: http://www.ashgate.com

British Library Cataloguing in Publication Data
Orford, Scott
    Valuing the built environment : GIS and house price
    analysis
    1. Housing - Prices - Great Britain
    I. Title
    333.3'3823'0941

Library of Congress Catalog Card Number: 99-72976

ISBN 0 7546 1012 8

Printed and bound by Athenaeum Press, Ltd.,
Gateshead, Tyne & Wear.

Contents

List of Figures

List of Tables

Preface

In the autumn of 1998, a group of residents in the Bristol suburb of Clifton formed a committee to campaign against a local development. This was to convert a wooded piece of common land into car parking space. The ownership of this land had historically been vague, and it had subsequently been adopted by the local residents as a communal garden. However, one of the residents suddenly announced that he had claimed 'squatters rights' on the land and proposed to convert the garden into valuable car parking space. Since such communal gardens are common in Clifton, and car parking space scarce, the fight between the campaign committee and the developer was watched with interest. Obviously, the loss of somewhere to sit during the summer was only of a minor concern. As one resident on the committee remarked:

> "I don't want to look out of my bedroom window and see cars and breathe exhaust fumes ... it will devalue our property prices."

This statement is the essence of the book. Local externalities, such as wooded areas of land and car parks, are inherent features of the built environment. They can simultaneously provide utility and disutility to local inhabitants and can affect property prices adversely. The experience in Clifton is only one of many that are occurring in towns and cities within the UK and beyond. It is a classic example of NIMBY-ism, a phenomenon that has become increasingly prevalent as Britain has become a Nation of Homeowners. There is now increasing evidence that this may intensify as the trend for urban living among the young and policies directed towards the development of brownfield sites turns the attention of developers and speculative builders back to established urban centres.

The Urban Studies literature contains numerous examples of NIMBY-ism, although these have usually been couched within social and political contexts. Beyond individual case studies, the economic implications of locational externalities remain rather vague. In particular, the precise nature of how locational externalities are incorporated into house prices is still unknown. This is despite the fact that their perceived effect upon house prices tends to be the very focus of conflict, as opposed to their immediate utility / disutility.

This book attempts to 'move towards a valuation of locational externalities' by providing a detailed investigation of how locational externalities are valued through house prices. It integrates the micro-economic theories of housing markets and residential location with the spatial statistical modelling literature. The result is a re-evaluation of the spatial dynamics of housing markets within a geographical as opposed to a purely economic context. As for the situation in Clifton, this has recently been resolved. It seems that the local residents will continue to enjoy drinking Pimms in their communal garden for many years to come.

Scott Orford   *Clifton, 1999*

Acknowledgements

There are numerous people whom I would like to thank in the School of Geographical Sciences in Bristol. In particular, Paul Longley who has advised and supported me throughout this research. I would also like to thank Edward Thomas for his computer expertise and constant support. I would also like to acknowledge the help of Gary Higgs in the construction of the GIS, and specifically in the manual matching of ADDRESS-POINT to the various datasets. I would also like to thank Simon Godden for help with the art work. I am also indebted to the numerous Estate Agents and Building Societies for supplying me with the necessary information free of charge.

List of Abbreviations

Structural Variables

Floor Area	Total Floor Area (sq-ft)
ET	End-Terraced Dwelling
MT	Mid-Terraced Dwelling
SD	Semi-Detached Dwelling
D	Detached Dwelling
FCB	Flats in Converted Building
FPB	Purpose Built Flats
M	Maisonette
B	Bungalow
EL	End-Link Dwelling
ML	Mid-Link Dwelling
Beds	Number of Bedrooms
Recs	Number of Recreation rooms
Baths	Number of Bathrooms
Showers	Number of Shower rooms
Full CH	Full Central Heating
Gas	Gas Central Heating
Garages	Number of Garages
ORP	Off-Road Parking
New	Age: New
Post 1964	Age: Post 1964
1918-64	Age: 1918 - 1964
Pre-1918	Age: Pre-1918
Gdn: None	Garden: None
Gdn: < 5m	Garden: Less than 5 metres
Gdn: 5-50m	Garden: 5 - 50 metres
Gdn: > 50m	Garden: More than 50 metres
Needs Mods	In need of modernisation
Swm Pool	Swimming Pool
Con	Conservatory

Locational Variables

Dist CBD	Accessibility to CBD
Dist MWAY	Accessibility to M4 motorway
Dist Station	Accessibility to railway stations
Hospitals	Proximity to hospitals
Sports	Proximity to sports centres
Community	Proximity to community centres
Institutional	Proximity to institutional centres
Shops	Proximity to local shops
Primary	Proximity to primary schools
Secondary	Proximity to secondary schools
Bute Park	Proximity to Bute Park
Parks	Proximity to parks / open space
Light Ind	Proximity to light industrial land-use
Heavy Ind	Proximity to heavy industrial land-use
Rail 0-50m	Rail 0 -50m
Rail 50-100m	Rail 50 - 100m
Rail 100-150m	Rail 100 - 150m
Rail 150-200m	Rail 150 - 200m
River 0-50m	River 0 - 50m
River 50-100m	River 50 - 100m
River 100-150m	River 100 - 150m
River 150-200m	River 150 - 200m
Primary Road	Road Type: Primary
Secondary Road	Road Type: Secondary
Residential Road	Road Type: Residential
Close	Road Type: Cul-de-sac / Close
Poor 0-50m	Street quality 0-50m: Poor
Below Ave 0-50m	Street quality 0-50m: Below Average
Above Ave 0-50m	Street quality 0-50m: Above Average
Good 0-50m	Street quality 0-50m: Good
Poor 50-100m	Street quality 50-100m: Poor
Below Ave 50-100m	Street quality 50-100m: Below Average
Above Ave 50-100m	Street quality 50-100m: Above Average
Good 50-100m	Street quality 50-100m: Good
Poor 100-200m	Street quality 100-200m: Poor
Below Ave 100-200m	Street quality 100-200m: Below Average
Above Ave 100-200m	Street quality 100-200m: Above Average
Good 100-200m	Street quality 100-200m: Good

Non-res Buildings	Street non-residential buildings
Sch: Willows	Sch Catchment: Willows High School
Sch: Fitzalan	Sch Catchment: Fitzalan High School
Sch: Cantonia	Sch Catchment: Cantonia High School
Sch: Cathays	Sch Catchment: Cathays High School
Sch: St Teilo's	Sch Catchment: St Teilo's High School
La > 50%	Percentage Local Authority tenure
% Open Space	Percentage of open space
% Non-Residential	Percentage of non-residential land-use
Density	Housing density
Q.Shop	Quality of local shops
Q.Transport	Quality of local public transport
Q.Sport	Quality of local sport facilities
Q.Parks	Quality of local parks
Q.Commuinty	Quality of local community facilities
H.Qual	Housing Quality
Social	Socio-economic class

Other Abbreviations

CHCS	Cardiff Housing Condition Survey
CPD	Central Postcode Directory
ED	Enumeration District
GIS	Geographic Information Systems
HCS Area	Housing Condition Survey Area
OSAPR	Ordnance Survey ADDRESS-POINT Reference
PAF	Postcode Address File
VIF	Variance Inflation Factor

1 Introduction

It is often said that the most important determinant of property prices is location, location, location. Hence the valuation of the built environment has long been a traditional concern of geographers, economists and planners. Over the years numerous conceptual, theoretical and empirical studies have attempted to formulate, model and quantify how the built environment is valued by its inhabitants. This book aims to complement and extend some of this work by investigating how the built environment is valued via the mechanisms of the urban housing market. More specifically, the aim of the book is to move towards a valuation of locational externalities through the modelling of housing market dynamics.

Locational externalities are becoming an increasingly significant component of local housing market economies, particularly as gentrification and urban regeneration continue to change the urban landscape. The pressures being placed upon the redevelopment of brownfield sites is resulting in town cramming and the loss of open space. Recent evidence is now suggesting that over-development and scavenging for land is encouraging a switch from NIMBY-ism (not in my backyard) to BIMBY-ism (bungalow in my back yard) (Evans, 1991), confusing the traditional debate of how local residents act to exclude unwanted (usually negative) externalities from their neighbourhood. What is emerging from these changes to the built environment is the need for a clearer understanding of how landuse impacts upon house prices. More importantly, there is a growing need to understand how these types of changes will ultimately be incorporated into the local housing market.

Empirical research relating to affects of externalities within a housing market is quite scarce. Work that does exist tends to be obscure and quite often either contradictory or counter-intuitive. To overcome this anomaly, this book intends to draw together existing work on locational externalities into a common structure based around the mechanisms of the urban housing market. In particular, the book aims to explore how housing market dynamics and locational externalities can be modelled using the method of hedonic pricing.

Hedonic pricing is an econometric technique used for estimating the monetary value of attributes of complex commodities. It is based upon the principles of pleasure seeking and consumer sovereignty; the belief that individuals are the best judges of their own needs and will maximise the set of attributes of a commodity that give them the most satisfaction. These attributes do not have directly observable market prices. Instead, the sum of their values is equivalent to the market price of the commodity. Within this context, the price of a house can be regarded as the sum of the implicit price of its specific bundle of attributes. Since location is an integral attribute of a house, its value within a bundle can be estimated using a hedonic house price function. This function relates house price to housing attributes, with the resulting parameter estimates corresponding to the implicit prices of these attributes.

Hedonic house price research has a long, established history, with its origins within the location and landuse theories of the 1960s. Although it has since moved away from the central tenants of these theories, it still continues to be intimately bound up with the micro-economic literature of housing markets and residential location. Moreover, the treatment of location within hedonic house price research often reflects the naïve treatment of location within these theories. It is therefore hardly surprising that previous work on locational externalities is somewhat problematic. Nevertheless, an understanding of housing market economics and residential location is necessary to comprehend how hedonic pricing can be used to value the built environment. The remainder of this chapter is devoted to a review of this literature and the extent to which hedonic house price theory has since departed from its original formulations. The review is necessarily discursive since the economic framework of hedonic house price theory is explained in detail in subsequent chapters.

Housing and the Housing Market

Housing as a Commodity

> 'It is fixed in geographic space, it changes hands infrequently, it is a commodity which we cannot do without, and it is a form of stored wealth which is subject to speculative activities in the market ... In addition, [it] has various forms of value to the user and above all it is the point from which the user relates to every other aspect of the urban scene' (Harvey, 1972; pp 16).

Housing is unlike most other commodities. It is a complex package of goods and services that extends well beyond the shelter provided by the dwelling itself. Housing is also a primary determinant of personal security, autonomy, comfort, well being and status, and the ownership of housing itself structures access to other scarce resources, such as educational, medical, financial and leisure facilities (Knox, 1995). As such, housing has been viewed as a 'composite demand for a flow of services embodying a variable mix of characteristics' (Maclennan, 1982. pp. 41) - a multidimensional commodity. It is so intimately bound up with the lives of individuals that only one is usually consumed by a household at any time.

It is typical to talk of a household purchasing packages or bundles of housing services that vary between housing types and housing markets. However, defining what a particular bundle actually is can be complicated. In addition, housing has a number of relatively unique attributes. It has a fixed location, a long durability, and a limited adaptability in response to changing demands. Housing stock is complex and diverse and is sensitive to changes that are external to the local market. Housing is also subjected to a multitude of institutional regulations imposed by government.

Supply and Demand of Housing

When discussing issues of supply and demand of housing, it is important to make the distinction between use value and exchange value. Use value generally refers to the net utility supplied by the bundle of housing services, whilst the exchange value is the capital value a property can realise in a competitive housing market. Although the use value of a property is a major determinant of its exchange value, this will also be influenced by the property's potential for increasing capital gain, since the purchase of housing stock is often the largest and only source of a households accumulated savings (Muth & Goodman, 1989). Thus, two different housing markets can be identified. One deals with the supply and demand of bundles of housing services, whilst the other deals with an asset that can be thought of as housing stock. Although these two markets are conceptually different, they are integrated, since the majority of houses will offer similar services, such as a water supply.

Therefore, housing demand is a reflection of both its use value for consumption or occupancy purposes and its exchange value as an investment good. Housing demand tends to vary between income and racial groups and at different stages of the family life cycle. Other influential factors include migration, immigration and changes in tax rates and taxation policies, particularly mortgage interest rates. Demand for housing has been of central interest to economic theorists and has been contextualised in micro-economic models of land use (Alonso, 1964) and residential location (Evans, 1973). However, these demand led models have been severely criticised for ignoring the supply of housing.

The majority of properties supplied on the housing market come from the existing housing stock. Only a small proportion of the supply comes from newly constructed properties (Bourne, 1981). Substantial proportions of new supplies from existing stock arise through the subdivision of property and the conversion of non-residential buildings to dwelling uses. Even more occur through the death of a household, through the move of an existing household to shared accommodation or by a move outside the city. Supplies of housing stock may be ended by demolition or conversion to a non-residential use, or even merger by knocking together two or more dwellings (Knox, 1995). Therefore, housing supply is a complex phenomenon with new supply and existing supply requiring separate, but interdependent analysis (Maclennan, 1982). Furthermore, the supply of housing will experience time lags between the decision to supply housing services and these housing services coming onto the market. The rate at which these new supplies enter the market in the short term is sensitive to house price changes and fluctuations in interest rates. This is particularly so for new constructions in which the price and the availability of land, planning controls and the provision of infrastructure have important influences on decisions regarding the location and timing of a development (Muth & Goodman, 1989).

Factors Influencing the Decisions to Move

The household is initially assumed to be receiving a given utility in their present dwelling. For movement to be considered, a minimum threshold of housing dissatisfaction must be perceived. This may occur over a period of time as the household recognises its existing mismatch of housing attributes and household activities. The decision to enter the housing market and evaluate alternative housing opportunities may be triggered by a variety of factors. These may include increases in income, changes in family size, household formulation, or relative price changes across the market. Knox (1995) has made the distinction between voluntary and involuntary moves. Voluntary movements may be initiated by dissatisfaction with dwelling and garden space, housing repair costs and style obsolescence, as well as complaints about the neighbourhood. Reasons for forced moves include marriage, divorce, a death in the family, retirement, ill health, and employment changes. However, almost two thirds of household movement is due to changes in the family life-cycle, and their perceived space requirements (Short, 1982).

In recent years another factor has emerged that has had an important influence on the propensity to move in the UK; negative equity. This occurs when the market price of a house becomes less than the mortgage secured upon it. This became a widespread problem at the end of the 1980s and during the 1990s when house prices slumped dramatically in many parts of the UK (Dorling, 1995). Negative equity means that the household is liable to cover the additional money secured on the property when it is sold, and having to find this money prevented many households from being able to move in the early 1990s. Negative equity particularly affected first time buyers, young buyers and less affluent buyers, since these were more likely to take out relatively larger loans and then have less ability to pay them back via earnings, inheritance and other assets. Low levels of equity can also deter households from moving.

The Micro-Economic Theory of Housing Markets

Introduction

The 1960s and 1970s witnessed a proliferation of micro-economic theories and mathematical models of housing markets, residential location and landuse in both the UK and USA (e.g. Alonso, 1964; Muth, 1969; Batty, 1976). These formulated housing market dynamics in purely economic terms, based upon the theory of the firm and consumer behaviour. A key element to these formulations was the concept of a housing market in perfect equilibrium functioning under Pareto Optimum conditions. These micro-economic theories subsequently underpinned neo-classical approaches to residential location and in particular, the trade-off model of residential location. The trade-off model is perhaps the most influential economic model of residential location within hedonic house price theory. However, as will be discussed, hedonic house price theory has since abandoned much of the micro-economic theory concerning perfectly functioning housing markets and Pareto Optimum conditionality, in favour of a segmented housing market in disequilibrium.

The Perfectly Competitive Housing Market

The owner-occupied housing market is primarily an economic market set within a political framework for the purpose of exchanging housing services. In economic theory, the role of the market is to allocate scarce resources in an efficient manner so as to maximise output while minimising cost, using price as the allocation mechanism. The most precise interpretations of this conceptualisation of the housing market derive primarily from the micro-economics literature. Maclennan (1982) identifies nine assumptions that define a set of conditions sufficient for the existence of a perfectly competitive housing market. These focus on the behaviour of individual producers and consumers, and regard the matching of households to housing units as essentially an assignment problem. The allocation proceeds as to achieve a market clearing solution; one in which all housing units are allocated and all households are accommodated in the most efficient way. The assignment is also optimal in the sense that no household could be made better off with a different assignment without making another household worse off. This is known as Pareto Optimum conditions, and is a source of contention in the hedonic house price literature that argues against Pareto Optimum equilibrium conditions in favour of a segmented housing market in disequilibrium. However, before this can be explored in more detail, it is necessary to set out the conditions under which a perfectly functioning housing market is said to operate.

Following Maclennan (1982, pp. 36), Pareto Optimum conditions can be achieved under the following nine assumptions:

1.   There are many buyers and sellers.
2.   In relation to the aggregate volume of transactions the sales or purchases of each house are insignificant.
3.   There is no collusion amongst or between buyers and sellers.
4.   There is free entry into and exit from the market for both consumers and producers.
5.   Consumers have continuous, transitive and established preferences over a wide range of alternative choices of housing and non-housing goods.
6.   Consumers and producers possess both perfect knowledge with respect to prevailing prices and current bids and perfect foresight with respect to future prices and future bids.
7.   Consumers maximise total utility1 whilst producers maximise total profits.
8.   There are no artificial restrictions placed on the demands for supplies and prices of housing services and the resources used to produce housing service. For instance, house purchases are not constrained by finance rationing or the non-availability of preferred housing choices.
9.   The market is assumed to be in equilibrium.

It can be seen that these assumptions are extremely idealised, and as such, easily critiqued. For instance, the abstracted assumptions of perfect competition and rational buyers and sellers are frequently cited (Ball, 1985). However, a major source of criticism of the micro-economic theories of housing markets has been the disregard of the supply of housing. Compared to demand, very little micro-economic work has been done on the supply of housing, especially in the short-run (Muth and Goodman, 1989). For instance, in the micro-economic supply model developed by Muth (1969), a supplier has perfect information regarding present and future house price changes, and is assumed to be a price-taking profit maximiser. This allows the precise output level of housing to be identified deductively. This model has been used to formulate theoretical specifications for the estimation of price elasticity of supply of housing both in the short-run and long-run and for the elasticity of substitution between land and non-land inputs to housing supply. However, the durability of housing, and the difficulty of adapting existing stock to changes in demand has been ignored, even though these will effect long- and short-run housing market equilibriums.

The Neo-Classical Approach to Residential Location

Introduction Concurrent with the formulation of the micro-economic theories of housing markets was the development of new approaches to residential location. Under the auspicious title of new urban economics, these neo-classical approaches were underpinned by similar micro-economic assumptions, and used comparative-static utility maximisation to deduced urban rent gradients and individual household demand functions for housing space and city centre access. By the mid-1970s these models had coalesced into a general theory of residential location known as the trade-off model.

The trade-off model The trade-off model (Basset & Short, 1980), or 'access-space' trade-off model (Maclennan, 1982), describes how households trade-off travel costs to the city centre, against housing costs in an attempt to maximise utility subject to an overall budget constraint. The theoretical background of the trade-off model was developed in two stages (Anas & Dendrinos, 1976). The basic models were developed during the 1960s, principally by Alonso (1964), Beckmann (1968) and Muth (1969), and were based upon the micro-economic theory of housing markets operating under Pareto Optimum conditions. These initial models were elaborated during the 1970s by economists such as Evans (1973), Mills

(1972) and MacDonald (1979). The principal contributions of these economists were an addition of a commuting and leisure time constraint in the household utility function (e.g. MacDonald, 1979), an interest in polio-centric urban forms (e.g. Evans, 1973), and the influence of neighbourhood (e.g. Papageorgiou, 1976). However, the fundamental principles still remained the same (Ball, 1985).

The trade-off model was developed under the assumptions of a monocentric city on an isotropic transport plane with a housing market in perfect competition. The basic premise of the model was that house size and access to the city centre were both important determinants of household utility (Muth & Goodman, 1989). Given that transport costs increased from the city centre at a diminishing rate and households always maximise their utility, the model deduced that land prices would fall at a decreasing rate as transport costs rose. Alonso (1964), who pioneered the model, regarded residential location as simply a conflict between spacious living and easy access to the city centre. In other words, how a household balances 'the costs and bother of commuting against the advantages of cheaper land with increasing distance from the center of the city and the satisfaction of more space' (pp. 15). Hence, the optimal location for a household was one where the decrease in housing costs with a move away from the city centre was equal to the increase in transport costs. A stable housing market equilibrium was eventually reached by each household choosing their optimal location through a bid-rent function.

The Bid-Rent function This was developed by Alonso (1964) and describes how a housing market equilibrium can be derived from the individual location demand functions of the trade-off model. If a household locates by trading off travel costs against housing costs, then associated with this trade off is a particular level of utility that is fixed equal to the maximum utility attainable if the household located at the city centre. To locate away from the city centre, the household needs to be indifferent with the new location if the same level of utility is to be maintained. This is achieved by the household bidding or stating the level of rent per unit housing they are prepared to pay at this new location. Since this price must be low enough to offset transport costs, housing costs decline with distance from the city centre. A bid-rent schedule can be obtained that indicates the relative priorities for rent and travel costs. This bid-rent schedule can be used to calculate a bid-rent curve, which describes bid-rents as a continuous function of distance, whilst holding utility constant. Each household has a complete set of bid-rent curves covering all possible rents and distances.

The lower the bid-rent curve, the lower the rent per unit housing and hence the higher the utility. (Muth & Goodman, 1989)

If all the households in the city have the same incomes, tastes and preferences, their set of bid-rent curves will be identical. If the city is in market equilibrium, the rent gradient will lie wholly along one of the bid-rent curves. If it does not, then households would adjust their location in order to maximise their utility until the rent gradient and the bid-rent curve coincide. The particular bid-rent curve that coincides with the rent gradient depends upon residential and non-residential demands for space. However, since households generally have different incomes, tastes and preferences, their set of bid-rent curves will not be identical. In this case, the rent gradient will not lie along a single bid-rent curve, but will be made up of sections of the lowest attainable bid-rent curves of all the households in the city. Hence, a household's optimal location will be the point at which the rent gradient is tangential to their lowest achievable bid-rent curve. At this point, the slope of the rent gradient is equal to the slope of their bid-rent curve and the household's utility is at its maximum.

Criticisms of the trade-off model The trade-off model has become a paradigm for much urban economic research. Its great strength lies in its heuristic power in producing results consistent with data from real cities and with findings of earlier urban theory (Evans, 1973; Maclennan, 1982; Muth & Goodman, 1989). In particular, the trade-off model deduces the existence of a negative rent gradient from the city centre outwards, which decreases with increasing distance. It is this feature of the trade-off model that has been the most influential in hedonic house price studies. One of the motivations behind early hedonic house price research was to estimate this negative rent gradient, as this would strengthen the argument for the concept of the bid-rent function.

Criticisms of the trade-off model are plentiful (Basset and Short, 1980; Maclennan, 1982). One of the main criticisms is that the trade-off model is demand orientated, with no regard for the supply of housing. With respect to the existing stock, supply has either been ignored or effortlessly adapted to variations in demand, 'almost in the fashion that children build with lego' (Bourne, 1981. pp. 131). Other substantive criticisms concern the fundamental importance of accessibility to the city centre in determining residential location. Other housing attributes, such as housing quality and population density, tend to be broadly correlated with distance from the city centre, and these may have more of an influence than issues of accessibility. A final criticism is that, whilst neo-classical models are relatively successful at describing residential location patterns, they fail to adequately explain them since they ignore the wider social structures and institutions that govern household decisions. They also suffer from a neglect of the social relations of housing provision in a historically specific context. As Ball (1985) observes, neo-classical models commit 'considerable violence to [our] common-sense understanding of urban spatial structures' (pp. 506).

Housing Market Disequilibrium and Segmentation

Introduction

The previous two sections have briefly described some conceptual and economic considerations that have underpinned hedonic house price theory. In particular, the trade-off model of residential location, and the deduction of a negative rent gradient from the city centre outwards, have been important theoretical constructs that have shaped much hedonic house price research. Indeed, the motivation behind the early hedonic house price research was to provide empirical evidence of a negative rent gradient as verification of the trade-off model. However, as will be discussed in subsequent chapters, empirical results generated by early hedonic research were inconsistent and contradictory, particularly in the estimation of the negative rent gradient. Moreover, the assumptions of a housing market in perfect equilibrium, operating under Pareto Optimum conditions were questioned. Instead the housing market was re-formulated in terms of segmentation and disequilibrium, with the imperfect knowledge of buyers and sellers, and the influence of institutions and actors in structuring the housing market becoming important. This has resulted in an interest in the concept of housing submarkets.

Housing Submarkets

> 'Heterogeneity in the existing stock, other differences in neighbourhood desirability, and the existence of discrimination imply that the urban housing market is a set of compartmentalised and unique submarkets delineated by housing type and location' (Schnare & Struyk, 1976; pp. 147).

Housing market disequilibrium occurs when changes in demand and supply are unequal. This may occur for a number of reasons. Prospective buyers often have limited house search areas due to search costs, imperfect costs, imperfect information, or a desire to be close to workplace, friends or relations. This will mean that only a limited number of housing bundles will be taken into consideration, which can lead to imperfect competition. There may also be highly inelastic demands for certain housing, especially in high quality neighbourhoods, and this could be confounded if the demand is shared by a large number of households. The very nature of housing means that supply is generally inelastic in the short run, and is quite often inelastic for some housing bundles over longer periods due to durability of stock which is difficult to modify, and a lack of building land constraining location. This usually means that housing demand will change more rapidly than its supply, and this is exacerbated by the time lag in new completions and conversions. The resulting disequilibrium may be pervasive and it will be compounded by investment decisions, causing a high degree of under-occupation and inefficient use of the housing stock (Short, 1982).

Moreover, restrictive supply and demand processes may segment the market into a number of more or less independent sectors, with local supply and demand mechanisms resulting in a different structure of prices in each. These sectors can be viewed in two domains: whether the stock is partitioned into distinct sectors in aspatial terms, or whether the urban area is also geographically subdivided into 'spatial submarkets' (Bourne, 1981). Most commentators now agree that a functional urban housing market does not operate as one large market, but rather as a series of linked, quasi-independent submarkets. Their existence is reflected by significant differences in prices paid for a given amount of housing services. Housing submarkets arise for several reasons. Firstly, they are the result of housing market disequilibrium caused by the factors discussed above. These factors will become exaggerated in larger urban areas through the sheer size and heterogeneity of the housing stock and the diversity of demands placed upon it by a more heterogeneous population. Secondly, they are the result of institutional barriers and are significantly influenced by the actions of gatekeepers such as land-owners, developers, estate agents, housing managers, and financial institutions whose motivation and behaviour largely structure the supply of housing (Knox, 1995). This is particularly important with respect to housing segmentation caused by racial discrimination.

Actors and Institutions in the Housing Market

Supply and demand opportunities are shaped and constrained by various agencies and professional mediators. These have been termed gatekeepers

(Saunders, 1990), and represent the institutions and agencies that operate at the interface between the housing stock and buyers and sellers. These include local government agencies, builders and landowners, although the two most documented examples of gatekeepers are mortgage lenders, such as building societies, and exchange professionals, such as estate agents.

Building societies have been documented (e.g. Boddy, 1980) to have a bias towards certain people, places and types of housing stock when allocating mortgages. In particular, people of colour, those on low incomes or part time employment, and old, large housing in deprived neighbourhoods are less likely to be granted a mortgage. Moreover, this may be translated into a spatial bias, with financial institutions avoiding what they regard as 'risky' areas. This is known as redlining, and is the reluctance to advance funds on any property within neighbourhoods perceived to be a bad risk, usually innercity areas with a high percentage of ethnic minority households and students. However, the effects of redlining are now of dwindling importance in the UK, given the changes in the provision of housing finance during the 1980s. Estate agents can also influence the allocation and distribution of housing in several ways. Since they control housing market information for both buyers and sellers, they may introduce bias by steering households into or away from specific markets.

Structure of the Book

This chapter has introduced the broad aims of the research and discussed some of the underlying themes. In particular, it has argued how hedonic house price theory has reformulated the micro-economic theories of location and landuse of the 1960s and 1970s to take into account the vagaries of the housing market. Specifically, the concept of submarkets has now become a central tenant of the economics of housing markets. Therefore, it can be concluded that in some respects, hedonic house price theory has better conceptualised housing market dynamics than conventional micro-economic theory. This shall now be expanded upon in chapter two, which is essentially a discussion of the hedonic house price function, its underlying theory and methodology. Of special importance are the ways in which space and location can be made an integral part of the model's specification. Chapter three continues this discussion but with respect to the specification of locational externalities and the generation of spatial data. I will argue that this has historically been poor, but with the advent of GIS technologies, new avenues of hedonic house price research have been opened. This is continued in chapter four, which describes in detail the construction of a context-sensitive urban GIS for the city of Cardiff. This, coupled with the availability of data geo-referenced to a high spatial resolution, has allowed locational externalities to be modelled at a detail not previously possible. The results are divided into two main parts. Chapter five investigates how the hedonic house price function can effectively model the dynamics of the Cardiff housing market. The main emphasis of this investigation concerns the spatial specification of the function and the diagnostic techniques that can ascertain how well the spatial structures of supply and demand are being modelled. The second results chapter, chapter six, uses the results of chapter five to move towards a valuation of locational externalities. The use of GIS in this chapter is perhaps its most important aspect, demonstrating how hedonic house price modelling and spatial analysis in GIS can be coupled to produce a geography of locational externality valuations. The final chapter develops an overview of the book and outlines several implications of the research from spatial econometrics to real estate valuations. The book concludes that house prices and housing market dynamics are inherently spatial, and hence that it is necessary to approach the valuation of the built environment from a geographical rather than a purely economic perspective.

2 The Hedonic House Price Function

Introduction

The hedonic house price function relates the price of a house to its attributes via the mechanisms of the housing market. Following certain assumptions, this makes possible the estimation of the implicit price of each housing attribute. These can then be used in the estimation of models of housing demand. The purpose of this chapter therefore will be to outline the main issues concerned with the estimation of the hedonic house price function. In particular, the specification of the function will be explored in detail since this is an area which I believe has been neglected and is poorly understood.

Utility Theory

When a commodity is consumed, some benefit or satisfaction is derived. This is called utility. Utility has been conceived as the property of an object that produces benefit, advantage, pleasure and happiness (Veldhuisen & Timmermans, 1984). It is based upon the principle of consumer sovereignty; the belief that individuals are the best judges of their own needs. A consumer will choose a commodity to gain the greatest benefit; to 'maximise his or her utility'. Utility theory identifies a consumer's utility function based on either assumed or revealed preferences and predicts choices constrained by the consumers level of income. Hence, following Freeman (1979a), the conventional utility maximisation problem may be expressed as:

maximise $U = U(X)$ 2.1
subject to $\Sigma (p_i x_i) = Y$

where U is a consumers utility function, X is a vector of commodities ($X = x_1, ..., x_n$), P is a vector of prices ($P = p_1, ..., p_n$), and Y is annual income. The solution to this problem leads to a set of ordinary demand functions conditional on prices and income, and some maximum utility level U_y.

$$x_i = x_i (P, Y) \qquad\qquad 2.2$$

This is shown graphically in Figure 2.1. Here, an indifference curve (U_2) joins together all the combinations of two commodities (x and y) which yield the same utility to the consumer. The slope of the curve is the marginal rate of substitution, and reveals the combinations of the two commodities to which the consumer is indifferent. To determine which combination is chosen, income levels and the price of the commodities also have to be taken into consideration. This is the budget constraint faced by the consumer, and is shown by the budget line in Figure 2.1. The highest point on the indifference curve that intersects the budget line is called the point of consumer equilibrium, and is the point at which the consumer is maximising his or her utility subject to their budget constraint (point 0 in Figure 2.1).

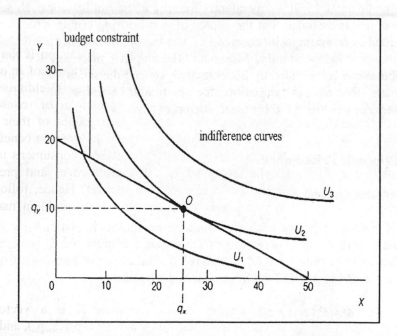

Figure 2.1 A hypothetical utility function for two commodities

Source: Johnston et al., (1994) pp. 277.

The 'Characteristics of Goods' Approach

There are many commodities that are not homogeneous but are traded on well-integrated markets. Houses and cars are both good examples. The utility provided by these commodities is based upon the utility yielded by their various attributes, rather than the composite good itself. The theory behind this 'characteristic of goods' approach was developed by Lancaster in 1966, and Griliches in 1971, and later expanded by Rosen (1974) who provided the theoretical framework for analysing a market for a single commodity with many attributes. Following Rosen, differentiated products like houses are assumed to be made up of bundles of attributes, which are not explicitly traded on the market, but as part of a package of housing services, Households are assumed to be utility maximisers and have a strongly separable utility function. A utility function is strongly separable if it can be partitioned into subsets, and the marginal rate of substitution between two commodities within a subset is independent of the quantities of commodities in other subsets. Consequently, households will divide their incomes between a subset of housing attributes and a subset of non-housing commodities, and will independently maximise their utility for each subset. Hence, if it is assumed that the vector of non-housing commodities can be regarded as one composite commodity, the bundle of housing attributes can be analysed independently. Moreover, the implicit prices of the housing attributes can be revealed by hedonic analysis and these can then be used to estimate the market valuation for particular housing attributes and subsequently the demand for these attributes.

The Hedonic Price Function

Theory and Overview

Let $Z = (z_1 \ldots z_n)$ be a vector of housing attributes. In Rosen's model of implicit markets, the interaction of supply and demand for Z produces a market clearing function P (Z) which relates the vector of housing attributes to the composite price of the house itself, such that:

$$P (Z) = P (z_1, \ldots, z_n) \qquad \qquad 2.3$$

P (Z) is the hedonic price function, and describes the house prices resulting from the interplay between housing supply and demand. Buyers and sellers take this price function as given in a competitive housing market. The P (Z) relationship between housing attributes and house prices need not be linear (Harrison & Rubinfeld, 1978). Non-linearities may exist because the housing market may not be in long-run equilibrium since housing supply is generally not very responsive to short term, and indeed long term changes in demand. Moreover, bundles of housing attributes cannot be untied and repackaged to reflect the consumers desired mix of housing services (Rosen, 1974. pp. 37-38). This is an important point. Whilst Lancaster (1966) assumed that the consumer could purchase each commodity in Z separately, Rosen argued that it is more reasonable to assume that the suppliers of housing sell bundles of housing services, Z, as part of a package. This has important implications in the specification of the hedonic price function.

Since the price of a property is a realisation of the price of its housing attributes, P (Z) can be estimated from observations of prices and attribute bundles of different houses. Moreover, the marginal implicit price of any attribute can be found by differentiating the hedonic price function with respect to that attribute. Hence:

$$\delta P (Z) / \delta (z_i) = P (z_i) \qquad\qquad 2.4$$

gives the increase in expenditure on Z that is required to obtain a house with one more unit of z_i, *ceteris paribus*. However, it is only under restrictive conditions that the function P (z_i) reflects the household demand for attribute z_i. The estimation of a household's marginal willingness to pay for an additional unit of z_i requires a further stage of analysis, and an understanding of the relationship between the marginal implicit price function, P (z_i), and housing market supply and demand functions.

Bid-Rent Functions and Marginal Willingness to Pay

Following Follain and Jimenez (1985), a household is assumed to have a strongly separable utility function U = U(X, Z), where X is the composite commodity of non-housing goods whose price is set equal to one, and Z is the vector of housing attributes. Households then maximise utility subject to the budget constraint Y = P (Z) + X where Y is the annual household income. The partial derivatives of the utility function with respect to a housing attribute are the household's marginal willingness to pay function for that attribute. In other words it represents the additional expenditure a consumer is willing to make on another unit of that attribute and be equally well off.

$$Uz_i / Ux = P (z_i) \equiv \delta P (Z) / \delta (z_i) \qquad 2.5$$
$i = 1,...,n$, under the usual properties of U

An important part of the Rosen model is the bid-rent function:

$$\theta = \theta (z_i, U, Y, \alpha) \qquad 2.6$$

where α is a parameter that differs from household to household (i.e. tastes). This can be characterised as the trade-off a household is willing to make between alternative quantities of a particular attribute at a given income and utility level, whilst remaining indifferent to the overall composition of consumption.

$$U = U (Y - \theta, Z , \alpha) \qquad 2.7$$

Tracing out these trade-offs generates a household's bid rent function for the given attribute, represented by θ^1 in the upper schedule of Figure 2.2. The household represented by θ^1 is everywhere indifferent along θ^1. θ schedules that are lower correspond to higher utility levels. At maximum utility, the bid-rent curve is tangential to the hedonic price function $P (Z)$. At this point:

$$\theta_i = Uz_i / Ux \qquad 2.8$$

which is the additional expenditure a consumer is willing to make on another unit of z_i and be equally well off (i.e. the demand curve). Figure 2.2 denotes two such equilibria: A for household θ^1 and B for household θ^2. However, since the hedonic price function represents the interplay between supply and demand, the supply side must also be considered. Since, like buyers, suppliers also accept $P (Z)$ as given, then the marginal cost of providing an attribute whilst maximising profits will be a concave offer curve ϕ that is tangential to $P (Z)$. Equilibrium points are those where supply equals demand. Since there are many consumers and many suppliers of housing attributes, there are many bid-rent and offer curves, and so $P (Z)$ represents a function consisting of the joint envelopes of various supply and demand tangencies (Muth & Goodman, 1989). This is shown in the lower schedule of Figure 2.2.

A household maximises utility by simultaneously moving along each marginal price schedule for a vector of housing attributes until it reaches a point where its marginal willingness to pay for an additional unit of each attribute just equals its marginal implicit price (Freeman, 1979b):

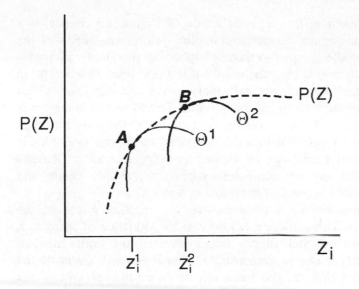

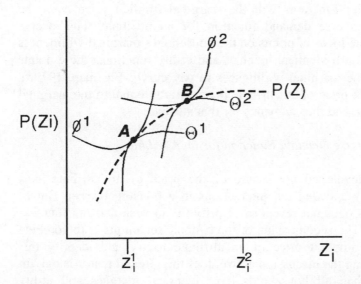

Figure 2.2 Demand and offer curves of the hedonic price function

Source: Follain and Jimenez, 1985; pp. 79; Fig. 1.

$$\theta_i = P(z_i) \qquad\qquad\qquad 2.9$$

This is shown in the upper schedule of Figure 2.3. Hence, if a household is in equilibrium, the marginal implicit prices associated with the chosen housing bundle is equal to the corresponding marginal willingness to pay for those attributes. Thus the marginal implicit price function of an attribute $P(z_i)$, is the locus of marginal supply curves and marginal bid-rent curves for that attribute by different households. This is shown in the lower schedule of Figure 2.3. Rosen's model is very similar to the standard urban (trade-off) model of residential location described in chapter one. This is because the trade-off model can be viewed as a special case of Rosen's model, that focuses on two attributes; access to the city centre and everything else about a house that generates utility.

The hedonic approach assumes that the implicit prices of the estimated hedonic price function, $P(z_i)$ reflects the valuation of attribute z_i as a result of demand and supply interactions of the entire market. However, in general it can be demonstrated that $P(z_i)$ will overstate the inverse demand function for the valuation of an additional unit of the attribute since, since to the right of points (a) and (b), $P(z_i) > \theta_i^1$ and θ_i^2 (see Figure 2.3). Only in extreme cases when all consumers have identical incomes and utility functions will the marginal implicit price curve be identical to the inverse demand function for an attribute. This occurs because $P(z_i)$ is the locus of points on the household's marginal willingness to pay curves θ_i. With identical incomes and utility functions, these points all fall on the same marginal willingness to pay curve (Freeman, 1979b). Hence, the implicit price of an attribute is not strictly equal to the marginal willingness-to-pay, and thus demands for that attribute.

Identifying the Inverse Demand Function for an Attribute

The above has developed a measure of the price of an attribute as a function of supply and demand interactions in a housing market. But as demonstrated, this does not reveal or identify the inverse demand function for that attribute. The second stage of the hedonic technique is to combine the quantity and implicit price of an attribute to try and identify this function. Expanding the discussion above does this. Recall that it is only in extreme cases when all households have identical incomes and utility functions that the marginal implicit price function is the same as its inverse demand function. By implication, by taking into account differences in household income, tastes and preferences and other household

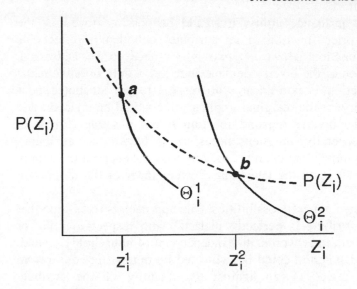

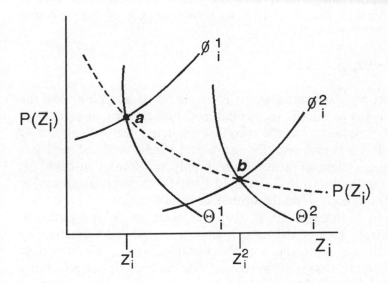

Figure 2.3 The marginal implicit price of an attribute as a function of supply and demand

Source: Follain and Jimenez, 1985; pp. 79; Fig. 1.

characteristics that influence utility, it should be possible to adjust the marginal implicit price function of an attribute such that it reflects its inverse demand function. However, two issues need to be addressed. Firstly, identification of the inverse demand function is only possible if the hedonic price function is non-linear with respect to the attribute under investigation. Otherwise the marginal implicit price would be constant and identification of the inverse demand function is not possible (Freeman, 1979b pp. 157). Secondly, the steps necessary to identify an attribute's demand function is dependent upon the assumptions made about the supply side of the market. It is on the latter that the remainder of the discussion will focus.

There are two general possibilities. One approach is to assume that the supply of an attribute is perfectly inelastic with respect to price or willingness to pay. In other words, it is independent of household demand. An inverse demand function could be estimated by regressing equilibrium marginal implicit prices $P(z_i)$ against the quantity of the attribute consumed, income levels and other variables. A second approach assumes that if both the quantities demanded and quantities supplied of an attribute is a function of price, then a simultaneous equation approach can be used to identify the demand function. This is known as the Rosen two-step approach.

The Rosen Two-Step Approach

This is the most popular method in the literature (Ohsfeldt, 1988), and the one which is closest to Rosen's theoretical model of the implicit market for characteristics. It requires that the marginal implicit price with respect to each attribute, $P(z_i)$, is evaluated for a particular bundle of Z and used as a price vector in a system of demand and supply equations that could be estimated simultaneously. Regression held constant for demand and supply shifts would theoretically yield the inverse demand function for a specified attribute (Follain & Jimenez, 1985). However, there are several problems with this approach. Two of the most important are possible simultaneity bias inherent in the approach, and the identification of the structural parameters (Ohsfeldt, 1988). Efforts to estimate the hedonic price function and to correct these problems has lead to innumerable serious statistical problems (Follain & Jimenez, 1985; Lerman and Kern, 1983; Gross, 1988; Graves et al 1988).

Assumptions of the Hedonic Price Function

There are several assumptions relating to the use of the hedonic price function as a basis for measuring the marginal implicit prices paid by households for bundles of housing attributes (Maclennan, 1982). These assumptions are of necessity very similar to those that underpin the workings of the competitive housing market although the housing market need not function as a unified whole, but maybe in disequilibrium. To summarise:

1. All consumers accurately perceive the attributes represented by the vector Z at every location.
2. There is sufficient variation in Z so that the hedonic price function P (Z) is continuous, with continuous first and second partial derivatives.
3. Spatial variations in housing attributes are capitalised into differentials in house prices.

Of course, any departure from these assumptions may invalidate the supposition that the hedonic price function can be used to estimate a household's valuation of housing attributes.

Hedonic Price Models and the Incorporation of Space

Introduction

The hedonic price model was developed within an economic framework to study essentially aspatial composite commodities such as cars, refrigerators, washing machines, and personal computers (Griliches, 1971; 1994). However, it has also been extensively used to analyse commodities, such as houses, which have spatial attributes. This has presented a fundamental problem; how to incorporate space into an aspatial econometric model. Neglecting the spatial element of these commodities can result in problems such as spatial heterogeneity, spatial autocorrelation, and multicollinearity that can produce distortions in the econometric model. These problems can also occur if the spatial element is only partially accounted for. This section is concerned with the evaluation of specifications of hedonic models, particularly with respect to their ability to handle spatial data.

The Traditional Specification

Housing attributes are the bundles of housing services that provide utility to the consumer. Fundamentally, Wilkinson (1973b) makes the distinction between dwelling specific or structural attributes and location specific attributes. The former are concerned with factors pertaining to the physical structure of a property, whilst the latter are concerned with the property's location. Hence, the hedonic price function can be defined as:

$$P(Z) = f(S,L) + \varepsilon \qquad\qquad 2.10$$

Where P is a vector of observed house prices, S and L are vectors of structural attributes and locational attributes respectively, and ε is a vector of random error terms. Typically, the specification of this function has been defined as:

$$P_i = \alpha\, X_i + \Sigma\, \beta_k\, S_{ki} + \Sigma\gamma_q\, L_{qi} + \varepsilon_i\, X_i \qquad\qquad 2.11$$

Where:
$i = 1, ..., N$ is the subscript denoting each property;
P_i is the price of property i;
$k = 1, ..., K$ is the number of structural attributes;
$q = 1, ..., Q$ is the number of locational attributes;
α, β, γ and ε are the corresponding parameters;
X_i is a column vector that consists entirely of ones.

This has been termed the traditional hedonic specification (Can, 1992), and has been the basic model in the majority of studies. If the attributes are taken as deviations from their mean, then the model suggests that the price of house i is a function of the average price of a typical property (α), the cost of structural and locational attributes (β_k) and (γ_q), and the price associated with the idiosyncratic elements of the individual house (ε_i). The model is estimated by ordinary least squares (OLS) regression, in which the regression coefficients represent the implicit price of each attribute. Hence, the hedonic price model has to satisfy the following OLS regression assumptions:

1. The relationship between the dependent variable (house price) and the independent variables (housing attributes) are linear in the parameters
2. The independent variables (housing attributes) are free from

multicollinearity
3. The errors terms are normally distributed with a mean of zero
4. The error terms are independent, that is, they are not autocorrelated
5. The error terms have a constant variance; that is, they are homoscedastic.

Any violation of these assumptions can lead to unreliable and biased parameter estimates. As will be seen, the violations of assumptions four and five are common feature of many studies, and are usually caused by the misspecification of the hedonic house price model. Misspecification issues have tended to be concerned with omitted variables and functional relationships. Hence:

'Finding the correct specification of the hedonic relationship for housing requires that we identify both the correct list of independent variables and the true functional form' (Butler, 1982; pp. 96).

These are important specification issues, since the wrong variables and an incorrect functional form can introduce bias into the model. Debates concerning the range and class of variables are discussed in detail later although at this stage it should be sufficient to note that theory offers little guidance in determining which particular attributes to include in the model (Ohsfeldt, 1988).

Functional Form

A fundamental issue in estimating the hedonic price function is choosing the functional form. A frequent criticism of hedonic studies is that functional form is chosen on the basis of convenience (Halvorsen & Pollakowski, 1982). Unfortunately, theory does not generally suggest a particular functional form for property attributes since, as Rosen has demonstrated, the hedonic price function is a reduced form equation reflecting both supply and demand mechanisms. Hence, the functional form may not be determined from information pertaining to either the underlying supply or demand equations. Instead its shape is determined by the distribution of housing bundle types and household types within a particular market area (Quigley, 1982). However, four functional forms are commonly used in hedonic house price models; linear, semi-log, log-linear and inverse semi-log (Palmquist, 1984).

It is important to impose a functional form that predetermines the correct relationship between the implicit price of a given attribute and the quantity of that attribute. For example, a linear functional form imposes the restriction that the implicit price of an attribute is constant across all quantities of that attribute. So, in the case of the implicit price of energy efficiency improvements (Johnson & Kaserman, 1983), an efficiency improvement in an extremely inefficient house is valued the same as an improvement in an extremely efficient house. Also, the use of a linear functional form requires the implicit price of energy efficiency to be independent of the level of other house attributes such as age and size (Dinan and Miranowski, 1989). More importantly, if the true functional form of the hedonic equation is not linear, the restriction of linearity may result in bias in the resulting coefficients (Linneman, 1980).

In the absence of theoretical guidelines, the Box-Cox, and occasionally the related Box-Tukey generalisation methods of transformations are often specified to search for an appropriate functional form (e.g. Freeman, 1979b; Halvorsen & Pollawski, 1981). Freeman (1979b) demonstrated that out of eight alternative hedonic price functions specified, only the Box-Cox transformation allows the implicit price of an attribute to depend upon the level of other attributes and to either decrease or increase as the level of the attribute varies. Although a full Box-Cox model may be specified, in which all the variables may take on a different power transformation factor, the usual procedure is to use a constrained version where all the continuous independent variables have the same power transformation factors, usually within a range of plausible values. However, there is no conceivable behavioural rationale for presuming this 'globally' imposed fit, with the only justification being one of minimising computational expense (Dunn et al, 1987). A full Box-Cox model may be specified as:

$$P\,(\theta) = \alpha + \Sigma\,\beta_k\,Z_k^{(\lambda_k)} + \mu \qquad\qquad\qquad 2.12$$

Where:
$$P\,(\theta) \;=\; \frac{P^{(\theta)} - 1}{\theta}\,, \qquad \theta \neq 0$$

$$=\; \ln P\,, \qquad \theta = 0$$

$$Z_k(\lambda k) \;=\; \frac{Z_k^{(\lambda k)} - 1}{\lambda_k}\,, \qquad \lambda_k \neq 0$$

$$=\; \ln Z_k\,, \qquad \lambda_k = 0$$

and Z is the vector of housing attributes. Then, in the constrained version, all $\lambda_k = \lambda$.

Furthermore, if the values of θ and λ are constrained equal to 1, the model reduces to the linear form. If θ and λ are constrained equal to 0, the model reduces to the log-linear form. If the value of θ is set equal to 0 and the value of λ is set equal to 1, then the semi-log model results. The opposite of the latter specification results in the inverse semi-log. Hence all the restricted functional forms commonly used are subcategories of the Box-Cox model, and statistical tests are used to determine which functional form best suits the data. But there is no reason why the testing of hypotheses should be 'straight-jacketed' into these most common functional forms (Longley & Dunn, 1988). It might also be difficult to choose between two or more specifications that have approximately the same scores on the statistical tests. Moreover, Dunn et al., (1987) have argued that the Box-Cox and Box-Tukey transformations are an undesirably mechanistic means of deriving functional form, and are unnecessarily clumsy and cumbersome in comparison to graphical diagnostic tests and exploratory data-analytic approaches in general. They also note that the Box-Cox and Box-Tukey transformations may not adequately account for the influence of outlying or anomalous data points, although this may be ameliorated by graphical techniques. By demonstrating how partial regression plots and other graphical diagnostics aided the derivation of the functional form of a logistic regression equation, they were able to conclude that such techniques offered a considerable improvement in flexibility and a greater coherence of interpretability compared to the restrictive Box-Cox and Box-Tukey traditions. However, such interactive data exploration techniques have been lacking in hedonic house price research, despite the introduction of user-friendly computer packages in recent years.

In terms of hedonic house price analysis, where the primary goal is to obtain accurate estimates of marginal prices, the functional form that generates the 'best fit' for the hedonic price function may not be the same as the functional form that generates the 'best' marginal price estimates (Ohlsfeldt, 1988). Indeed, in the work by Halvorsen & Pollakowski (1981), fewer counter-intuitive negative marginal price estimates were obtained using less complex functional forms. Similarly, in a simulation study by Cropper et al (1988), complex functional forms produced much greater errors in marginal price estimates. They both concluded that a simple linear Box-Cox specification of the hedonic price model generates the smallest errors in marginal price estimates. However, in light of the above discussion on the problems of using such mechanistic and restrictive methods, particularly with respect to unusual data points, such conclusions can be regarded as somewhat naive.

Spatial Misspecifications of the Traditional Hedonic Model

The traditional hedonic specification assumes that the effects of structural attributes on property values are fixed across the housing market, and hence each property will have the same marginal implicit prices. Locational attributes are incorporated as an additional set of housing attributes, independent of the structural attributes. This suggests that a household evaluates the structural attributes of a house, and the attributes of its location, separately. Can (1990) suggests that in this conceptualisation, location can be regarded as an additional premium on the price of a house, independent of the cost of the structural attributes. Furthermore, this specification suggests that there is no interaction or relationship between the structure of a house and its location within a city, which contradicts urban economic theory.

In recent years, it has become apparent that the traditional hedonic specification has not fully captured the spatial element of the data (Can, 1990; 1992). In particular, traditional hedonic models may suffer from spatial dependence and spatial heteroscedasticity. The problems caused by such spatial effects on the validity of traditional statistical methods has long been recognised (Anselin, 1988a). In particular, spatial effects will violate the assumptions of independently, identically distributed errors in the OLS regression model (assumptions four and five) used to estimate the hedonic model.

The problem of uncontrolled spatial effects in a hedonic model can be illustrated by the often-quoted study of the demand for clean air by Harrisons & Rubinfeld (1978). In this study, the Rosen two-step approach was used to estimate willingness-to-pay curves for air quality improvements. A traditional hedonic specification was estimated. heteroscedasticity was discovered and the model was subsequently re-estimated using weighted least squares. The model has since been re-analysed by Belsley et al., (1980). Diagnostic tests suggest that spatial autocorrelation is present and that the model may also suffered from spatial heteroscedasticity. If Belsley et. al., (1980) findings are correct, then it can be assumed that Harrisons & Rubinfeld's (1978) willingness-to-pay results are seriously flawed.

Therefore, the two spatial effects, spatial heterogeneity and spatial dependency, must be resolved if the multiple regression models used in evaluating the hedonic house prices are not to be invalidated. However, their effects in mainstream statistical and econometric literature have been almost totally ignored. As Anselin & Griffith (1988) have concluded:

> '[E]ven though the methodological results achieved in the fields of spatial statistics and spatial econometrics have been substantial, the dissemination from research community to applied world has been virtually non-existent' (pp. 14).

This is typical of most hedonic house price research, even though the data are likely have inherent spatial structures and be subject to various spill-over effects. This ignorance can be explained in part by the fact that the standard tests for functional misspecification, the selection of variables and the evaluation of predictive performance are not affected by the spatial nature of the models and data (Anselin, 1988a. pp. 282).

Spatial Heterogeneity and Housing Submarkets

> 'The central problem in estimating hedonic equations involves the delineation of homogeneous submarkets' (Straszheim, 1974; pp. 404).

The assumption that structural attributes will have the same fixed marginal implicit prices across urban space implies the presence of a single homogeneous competitive market. This fails to take into account the housing market dynamics that can lead to submarket formation. Generally, property prices have been conceived as varying continuously across urban space. Warnings against such a view, such as by Schnare & Struyk (1976), have generally been ignored. However, urban space is divided up into discrete units by transportation routes, housing stock and landuses. Spatial spill-over effects from these units imply that property prices are better conceived as contiguous rather than continuous. Since, by definition each of the submarkets will have a unique supply and demand structure, the implicit prices of the attributes will no longer be constant, but vary by submarket. If uncontrolled for, this spatial heterogeneity of implicit prices will cause structural instability in the regression coefficient and error term (Can, 1992). The result is a special case of heteroscedasticity that will violate the assumption of constant error variance.

Previous studies (e.g. Ball & Kirwin, 1977; Schnare & Struyk, 1976; Goodman, 1981) have tried to deal with spatial heterogeneity caused by the presence of submarkets by the method of 'switching regression' (Can, 1992). This involves estimating a hedonic house price model for the entire housing market and then separate ones for each submarket, with the specification of the model only concerned with the structural attributes of each house. Hence, if the housing market is divided up into 'M' discrete submarkets, then:

$$P_j (Z) = f_j (S) + \varepsilon_j \qquad j = 1, ..., M \qquad\qquad 2.13$$

Where P_j is the vector of house prices in submarket j
S is the vector of structural attributes.

If a statistically significant difference exists between the estimated coefficients for the entire market model and the coefficients in each of the submarket models, then this may provide evidence for structural instability and thus indicate the existence of a fragmented housing market.

Unfortunately, the method is highly complex and somewhat arbitrary. There are two main problems: identification and verification of potential submarkets. The method requires submarkets to be identified *a priori*, but this is problematic due to the hidden nature of the geography of supply and demand mechanisms that determine the implicit prices of the attributes. An indication of differential submarket mechanisms can be gained from an examination of house price inflation across the housing market. If there are varying demand and supply processes at work, then it can be expected that relative house price inflation will vary and houses in some areas will increase in price faster than others (Munro & Maclennan, 1987). In practice however, submarkets have generally been identified by either housing attributes or along neighbourhood definitions. The former has involved segmenting the dataset by the attributes of the housing stock, such as property type or tenure, whilst the latter by defined areas, typically based upon existing geographies such as political boundaries or census areas. A further stratification scheme has been to segment the data set by household characteristics such as race or socio-economic class, since these will influence the geography of the supply and demand schedules that are being defined. Schnare and Struyk (1976) suggest experimenting with several stratification schemes, using any large and significant differences in estimated parameters as evidence of submarket existence. Goodman (1981) suggests a more objective approach, using a method of defining submarkets based upon choosing the fewest number of submarkets possible, with the housing bundles within each being as similar as possible. He also suggests that since public services are an important part of a neighbourhood, houses within a given municipality should be placed in the same submarket. He argues that this should minimise the effect of differences in the cost and supply of public services between each municipality. Unlike Schnare and Struyk, Goodman (1981) suggests using analysis of covariance as verification of the submarket. He also points out that, since submarkets reflect differences in supply and demand schedules across the housing market as whole, the functional form of the hedonic model may also vary between submarkets.

The influence of submarkets has been contradictory. Whilst Goodman (1979; 1981) concluded that the implicit prices of attributes vary between submarkets, Schnare & Struyk (1976) and Ball & Kirwin (1977) discovered that the differences between the estimated coefficients for the submarket models and the model estimated for the whole housing market were insignificant. However urban housing markets are unique, and hence submarkets may form in some urban areas but not others, and this may explain the contradictory results. Furthermore, the arbitrary nature of defining submarkets by administrative geographies may lead to significant heterogeneity of supply and demand schedules within so called homogeneous submarkets. Also, the disaggregation of the housing market into discrete areas may be unrealistic, since the influence of certain locational attributes will extend beyond submarket boundaries. This is discussed in more detail in chapter three.

Spatial Dependence

The second problem is one of spatial dependency or spatial autocorrelation. Spatial autocorrelation will violate the assumption of independent errors. This could lead to misleading inferences about the significance of parameter estimates and can also negatively effect the validity of a wide range of standard diagnostics test. It occurs in regression analysis by two distinct forms of misspecification. Firstly, a variety of common misspecifications can result in spatial autocorrelation. Anselin (1988b) summaries these as factors associated with spatial aggregation, the presence of uncontrolled for non-linear relationships, and the omission of relevant variables. An example of the first is when aggregation of data results in spatial heterogeneity, such as treating the housing market as a unified whole instead of as a series of submarkets, and this produces spatial autocorrelation in the error term.

The second cause of spatial autocorrelation results when spatial data is incorrectly modelled. This is more fundamental in the sense that it is a special feature of spatial data (Can, 1990). It occurs in hedonic research since, firstly, the prices of nearby houses are similar because they share common locational attributes and will tend to have similar structural attributes, and secondly, because the prices of nearby houses will have an absolute or externality effect upon each other. The first consideration is the basis of house price theory, and will only be problematic with respect to omitted locational attributes. The second is less tangible. Can argued that the workings of the housing market were such that estate agents and buyers base the price of a house not only upon its structural and locational attributes, but also on the prices of properties in the immediate vicinity. In the same way, home owners may forego certain home improvements if they perceive that the affect on the capital value of the property will be minimal with respect to house prices in the immediate area. This has been corroborated by anecdotal evidence from estate agents who expressed the problems of selling housing that had been 'upgraded beyond the selling price of the area that it was located in'. Can described these as 'adjacency effects'.

A third complicating factor in model specification is the joint occurrence of both spatial effects, since factors which cause spatial autocorrelation are also likely to lead to spatial heterogeneity (Anselin, 1988a. pp. 290).

Contextual Hedonic Models

Introduction

The effects of spatial data in hedonic models have been generally ignored. This is despite the fact that such effects are pervasive and can be expected to be a fundamental feature of house price data. The traditional hedonic specification attempts to capture spatial variation solely by the use of locational attributes, which historically have been poorly specified - see chapter three. But the factors that contribute to spatial heterogeneity and spatial dependence are inherent in the structure of the data and the mechanisms of the housing market. Thus, Can (1990; 1992) has argued that the spatial aspects of house price data should be modelled explicitly within the specification. This requires an alternative specification of the hedonic function, that will have to take into account both submarkets (spatial heterogeneity) and the price of adjacent houses (spatial dependence). Such specifications can be regarded as 'contextual hedonic models' since they take into account the context of the housing market. Two types of contextual models have been proposed: spatial expansion hedonic models (Can, 1990;1992) and multi-level hedonic models (Jones & Bullen,

1993;1994).

The Spatial Expansion Specification

Introduction Can developed a series of hedonic model specifications to deal with spatial effects that are based upon the expansion method (Casetti, 1972. 1992). The expansion method is a technique for generating mathematical and statistical models by expanding the parameters of simpler models. It has become a well known procedure for 'contextualising' existing models (Foster, 1991), and has been used 'to ascertain how, where, when and why [functional] relationships vary from context to context' (Jones, 1991. pp. 45). Since this can include spatial context, Odland et al (1989) characterised the expansion method as a way to address specification errors arising from spatially heterogeneous processes, such as housing market dynamics. The expansion method inherently allows for the presence of spatial heterogeneity (submarkets), and can eliminate the part of heterogeneity resulting from spatial structural instability.

Parameter drift An important concept of the expansion method is 'parameter drift'. Parameters are said to 'drift' if their estimates significantly differ with context. Thus, parameter estimates of housing attributes can be thought to drift across submarkets. In the traditional hedonic specification, these parameters are assumed to be stable and invariant. '[The] bias is towards presupposing parameter stability, while the opposite should be true' (Cassetti, 1992: pp. 35). Furthermore, Cassetti argues that 'in the social sciences, functional relationships are likely to represent subsystems that will perform differently in different environments and circumstances rather than invariant laws' (ibid.). This suggests that spatial context can not be simply equated or controlled for through *ceteris paribus* conditionality. Interaction between the estimated parameters of a dwelling's structural attributes and its spatial location is needed if the contextual 'real world' is to be captured.

Using the expansion method to incorporate space Foster (1991) suggests that since any location in time and space is unique, reference to a location will provide a unique reference to a given context. 'Spatial' drift will only occur if the contextual variables are themselves systematically distributed through space. Otherwise, "contextual" drift is said to occur. Cassetti (1992) identified a three stage procedure for generating expanded models. In the first stage, an initial (global) model is generated, in this case the traditional hedonic model. Results from this model can be regarded as characteristic of the average context, which is uniformly applied to all observations. The second stage involves expanding some or all of the parameters in the initial model by equations that redefine them as functions of other variables. So in the case of the hedonic model, the structural attributes may be redefined as functions of locational attributes. These are called expansion equations. In the final stage, a terminal model is generated by placing the expansion parameters into the initial model. This is called an expanded model, or a spatially expanded model if the expansion equations contain a spatial element. In this way, space has been explicitly incorporated into the specification.

Expanding the hedonic price function Can (1990) argued that the structural parameters may take different values across urban space, varying with respect to location. She cites the example of a two car garage, which in the traditional specification will have the same marginal implicit price in an inner city neighbourhood, characterised by low car ownership, as in a suburb where the demand will be greater. Thus, using the three stage method, the hedonic price function may be expanded using the traditional hedonic specification as the initial model and location as the expansion equations. This is still problematic though, since the functional form and the attributes to be included in the expansion equations cannot be known *a priori*. Thus:

$$P\,(Z) = f\,(L\,(f\,(S))) + \varepsilon \qquad\qquad 2.14$$

where L is a measure of location and S are the structural attributes

This is the spatial expansion hedonic function. Since the spatial expansion method can be used to capture the spatial heterogeneity of submarkets, a typical expansion specification is one that accommodates discrete space. For instance, the intercept in the traditional hedonic model (equation 2.11) may be allowed to vary for M submarkets:

$$\alpha = \alpha_0 + \alpha_1 D_1 + \alpha_2 D_2 + ... + \alpha_{m-1} D_{m-1}$$
$$= \alpha_0 + \Sigma\, \alpha_j D_j \qquad\qquad 2.15$$

where j = 1, ..., M-1

The intercept term has been expanded with respect to a series of (M

- 1) dummy variables (D_j) that represent the individual submarkets. Each of the intercept terms ($\alpha_1 \ldots \alpha_{m-1}$) represents the submarket differentials from the base submarket price of α_0. If the expansion equation is placed into the initial equation (equation 2.11), the spatially expanded hedonic specification is as thus:

$$P_i = \Sigma (\alpha_0 + \alpha_j D_j) X_i + \Sigma \beta_k S_{ki} + \Sigma \gamma_q L_{qi} + \varepsilon_i X_i \qquad 2.16$$

The average house price varies between submarkets, but since the structural and locational attributes have not been expanded, the functional relationship between price and the housing attributes are invariant across space. To correct for this, the model may be re-specified such that the structural attributes, say, may also vary across submarkets.

$$\beta_k = \beta_{10} + \beta_{11} D_1 + \ldots + \beta_{20} + \beta_{21} D_1 + \ldots + \beta_{Kj-1} D_{j-1} + \beta_{Kj} D_j$$
$$= \beta_{k0} + \Sigma \beta_{kj} D_j \qquad 2.17$$

where: $k = 1, \ldots K$, and $j = 1, \ldots, M-1$

Substituting this into equation 2.16:

$$P_i = \Sigma (\alpha_0 + \alpha_j D_j) X_i + \Sigma (\beta_{k0} + \beta_{kj} D_j) S_{ki} + \Sigma \gamma_q L_{qi} + \varepsilon_i X_i \qquad 2.18$$

Not only does the average house price vary between submarkets, but also the marginal implicit prices of the structural attributes. The locational attributes can be expanded in a similar way.

$$P_i = \Sigma (\alpha_0 + \alpha_j D_j) X_i + \Sigma (\beta_{k0} + \beta_{kj} D_j) S_{ki} +$$
$$\Sigma (\gamma_{q0} + \gamma_{qj} D_j) L_{qi} + \varepsilon_i X_I \qquad 2.19$$

Jones & Bullen (1994) have pointed out that the result is equivalent to fitting a separate regression model between price and housing attributes for each submarket, which is very similar to the switching regression approach. Moreover, they stress that these two expansion models (equations 2.16 & 2.19) are nothing other than *ANOVA* and *ANCOVA* models. The major conceptual difference between the fully expanded model (2.19) and the switching regression model (2.13) is that the former has a single error term and thus assumes that the error variance is constant throughout the housing market. Since switching regression estimates separate models for each submarket, this assumption is relaxed. If this is the case, then the fully expanded model will be more efficient than the set of separate regressions. However, as Jones & Bullen (1994) argue, there may be no real gain in stringing together the separate regressions in the fully expanded model since the estimated coefficients will be identical, although the estimated standard errors of the coefficients may differ if the latter model is indeed more efficient.

The discrete spatial expansion methods above indicate that implicit prices vary with submarkets. An alternative expansion is to contextualise discrete space with locational attributes. Can (1992) operationalized this by constructing a measure of neighbourhood quality from census data and used this as a locational attribute. Two expansion equations were then specified, based upon a linear and quadratic functional form. The linear expansion equation of β can be specified as thus:

$$\beta_k = \beta_{10} + \beta_{11}NQ + \dots + \beta_{20} + \beta_{21}NQ + \dots + \beta_{K0}NQ + \beta_{K1}NQ$$
$$= \Sigma\,(\beta_{k0} + \beta_{k1}NQ) \qquad\qquad 2.20$$

where $k = 1, \dots K$ and NQ is a measure of neighbourhood quality

Hence, placing the expansion equation into the initial model:

$$P_i = \alpha\,X_i + \Sigma\,(\beta_{k0} + \beta_{k1}NQ)\,S_{ki} + \epsilon_i\,X_i \qquad\qquad 2.21$$

The terminal model (equation 2.21) asserts that location has no implicit price, but instead is seen as 'driving the spatial variation in the housing price determination process' (Can, 1990. pp. 258). This concept is supported by Witte et al, (1979), who have argued that implicit markets do not exist for locational attributes, since they are not determined by the action of any single supplier, but rather are the result of multiple independent decisions of the inhabitants. Instead, they serve to shift both the supply and demand curves of structural attributes, and hence the implicit price of the attributes. However, this is debatable, and it will be argued that the implicit markets do exist for locational attributes, and thus implicit prices, but not at the level of the individual house.

The specification incorporates the effects of spatial heterogeneity by allowing the structural attributes to 'drift' with neighbourhood quality, as measured by census data. However, this implies that the geography of the census tracts is a good proxy for submarkets, and moreover, that the spatial heterogeneity of submarkets can be adequately captured by neighbourhood quality differentials, and that the relationship between parameter drift and neighbourhood quality is consistent across the housing market. Also, since the intercept is left unexpanded, the model implies that the price of the average house is the same across the entire market, independent of neighbourhood context and submarkets. As such, Can (1992) concluded that although the spatial expansion hedonic specification reduced heteroscedasticity in comparison to the traditional hedonic specification, it was not completely removed.

Spatial dependence caused by the spill over effects of nearby houses cannot be ameliorated by the expansion method alone. Instead, Can (1992) proposed incorporating an autoregressive function into the spatial expansion specification.

$$P_i = \alpha\,X_i + \rho\,WP + \Sigma\,(\beta_{k0} + \beta_{k1}NQ)\,S_{ki} + \varepsilon_i\,X_i \qquad 2.22$$

Where W is the generalised weight matrix; WP is the spatially lagged dependent variable (house price), and ρ is its coefficient. The hypothesised spatial dependence is determined by W that is specified *a priori*. In this specification the price of a house is dependent upon the price of properties at nearby locations in addition to its structural and neighbourhood attributes. The coefficient ρ measures this absolute price effect of nearby properties. Since the specification contains a lagged dependent variable WP, OLS regression cannot be used, since the assumption of independent errors will be violated. In this case, OLS would lead to both biased and inconsistent parameter estimates. Can (1992) proposed using a maximum likelihood estimator, and concluded that hedonic models that contained a lagged dependent variable no longer suffered from spatial dependence.

The Multi-level Specification

Introduction The spatial expansion method incorporates contextuality into the specification by expanding what Jones (1991) has described as the 'fixed' parts of the hedonic model. These refer to the parameters of the attributes and the intercept term, which are 'fixed' or unchanging as opposed to the error term, or random part of the model that is taken from a distribution. Jones & Bullen have argued that fixed-part expansions are problematic:

> '[Neighbourhood] quality has been made an attribute of each housing unit, with no distinction between houses and the [submarkets] in which they are located. [Submarkets] and houses are treated as equivalent

observations, although houses are likely to be more numerous than [submarkets], and houses within a [submarket] are likely to be more similar than houses in a different [submarket]. When there is only one observation per [submarket], the within-place variation is totally confounded with the between-place variation and no separate estimates of these distinct components is possible.' (Jones & Bullen, 1994. pp. 255).

The argument is concerned with the differentiation between compositional and contextual effects (Jones & Bullen, 1993). The major problem is that the contextual effects, that is the difference a place makes, are potentially confounded with the compositional effects, or the differences produced by the variations in housing attributes within each place. Such was acknowledged early on by Wilkinson (1973a), who argued that '[i]t is difficult conceptually as well as statistically to distinguish the effects of the characteristics of a dwelling alone on price since an obvious and important feature of a neighbourhood is its stock of dwellings' (pp. 76). This has been the result of treating the location of a property in one dimension or spatial level.

Instead, Jones & Bullen (1993) conclude that, in this case, there are two distinct levels of analysis: houses (level 1) and submarkets (level 2), whereas the spatial expansion methods presume only one single level.

'The single level model assumes that the data does not have a hierarchical structure, that all the relevant variation is at one scale, that there is no auto-correlation and that there is a single general relationship across space and time ... [T]his model denies geography and history; everywhere and anytime is basically the same!' (Jones, 1991. pp. 8)

Inferential errors are likely to occur when inappropriate single-level models are used, and when multi-level data are modelled using techniques designed for a random sample, such as OLS regression. These problems can be overcome by specifying the model, not as varying at a single level, but as varying simultaneously over a number of levels. This is achieved by a modification of the three-stage expansion method. If the error term, or random part of the model is expanded as opposed to the fixed-part, then the expansion can be specified at a higher level. The results of such an expansion is known as a multi-level model (Goldstein, 1987).

Using the multi-level method to incorporate space After Jones & Bullen (1994), the traditional hedonic specification can now be written in terms of fixed and random parts, algebraically detailing the house and submarket levels. For the sake of clarity, locational and structural attributes have been reduced to a vector Z of K housing attributes.

$$P_{ij} = \alpha_j X_{ij} + \Sigma \beta_k Z_{kij} + \varepsilon_{ij} X_{ij}$$ 2.23

Where: $j = 1, ..., M$
$i = 1, ..., N$
$k = 1, ..., K$

where Z is the vector of K housing attributes, and M is the number of submarkets, and N is the number of properties. This is a micro-model, since it is based upon individual data. To achieve the equivalent multi-level model to that of the discrete-space fixed expansion of equation 2.16, the intercept has to be allowed to vary in a higher, level 2 between submarket, model, by the expansion equation:

$$\alpha_j = \alpha + \mu_{\alpha j}$$ 2.24

This is a macro-model, since it is based upon aggregated data. The price of the typical house in submarket j (α_j), is seen as a function of the market-wide price, α, plus a differential for each submarket, $\mu_{\alpha j}$. The micro-model is the within-place equation, whilst the macro-model is the between -place equation in which one of the parameters of the within-place model, in the case the intercept, is the dependent variable. Both models combines to form the terminal model:

$$P_{ij} = \alpha_j X_{ij} + \Sigma \beta_k Z_{kij} + (\mu_{\alpha j} X_{ij} + \varepsilon_{ij} X_{ij})$$ 2.25

The two random terms in the brackets are assumed to be independent of each other (Goldstein, 1987). Since the intercepts are allowed to vary according to a distribution, Jones (1991) termed these random intercepts models. Fully random multi-level hedonic models also allow the attribute parameters to vary according to a higher level distribution. This is achieved by specifying an additional macro-model:

$$\beta_{kj} = \beta_k + \mu_{\beta kj}$$ 2.26

This conceives the attribute's implicit price as an average market-wide price plus a submarket differential. The combination of the initial model and the two macro-models produces the terminal model:

$$P_{ij} = \alpha_j X_{ij} + \Sigma \beta_k Z_{kij} + (\mu_{\alpha j} X_{ij} + \mu_{\beta kj} Z_{kij} + \varepsilon_{ij} X_{ij}) \qquad 2.27$$

The model now has four random terms; σ^2_ε at level 1; and $\sigma^2_{\mu\alpha}$, $\sigma^2_{\mu\beta}$ and the covariance term, $\sigma_{\mu\alpha\mu\beta}$ at level 2. The covariance term allows the random intercepts and attribute parameters to co-vary according to a higher level, joint distribution. It is this concept of a higher level distribution which is the key to multi-level models (Goldstein, 1987). The differential intercepts and implicit prices are not specified as fixed, separate and independent as in the usual fixed-part expansion model, but as coming from a distribution at a higher level. Since these distributions concern not houses but submarkets, it is identical to treating places as a sample drawn from a population. The means of these higher level distributions are simply the usual intercept and implicit prices representing the average market-wide relationship. It is the variances / covariance's of the higher-level random terms that capture parameter drift. Moreover, if these variance terms are effectively zero, there is no parameter drift and there is no need for macro-models. There are no significant submarket effects and the traditional hedonic specification is adequate in describing house price variation.

Multi-level models and spatial effects Since the model contains more than one error term, it cannot be estimated using OLS regression. Instead, a procedure using an Iterative Generalised Least-Squares (IGLS) algorithm can be used (Goldstein, 1987; Jones, 1991). This algorithm simultaneously estimates the fixed and random parameters in a sequence of linear regressions until it reaches a convergence. An advantage of IGLS is that, unlike OLS, IGLS explicitly models spatial dependence and spatial heterogeneity.

   With multi-level data, such as houses nested in submarkets, spatial dependence can be treated as the norm since individual houses in the same submarket are likely to be more similar, in some way, than houses drawn from the entire housing market at random. Hence, autocorrelation is to be expected in hierarchical data, and the multi-level approach exploits this dependence to derive improved estimates, while the standard errors of the estimates are adjusted to take into account the autocorrelation (Goldstein, 1987). The degree of autocorrelation in multi-level models can loosely be conceived as the ratio of 'variation at the higher level' to the 'total variation of all levels'. If the ratio is zero, there is no autocorrelation and only a single level model is needed.

   With respect to spatial heterogeneity, multi-level modelling can be seen as a technique that models the structure of the variation that is not accounted for by the housing attributes, and does not assume a constant variance that can be captured by a single error term like OLS regression. Jones & Bullen (1994) illustrate this by demonstrating that the total variance for equation 2.27 between the submarkets at level 2 is the sum of the two random variables:

$$\text{Var} \left(\mu_{\alpha j} X_{ij} + \mu_{\beta k j} Z_{kij} \right) = \sigma^2_{\mu \alpha} X_{ij} + 2\sigma_{\mu \alpha \mu \beta} Z_{kij} + \sigma^2_{\mu \beta k} X_{ij}^2 \qquad 2.28$$

The total variance is not constant but a quadratic function of the attributes. In such a specification, the greatest variation between submarkets may be for large and small properties, say, with the smallest spatial variation in prices for properties of average size. Furthermore, multi-level modelling is capable of capturing heteroscedasticity caused by more common misspecifications, such as heteroscedasticity caused by an attribute that has been draw from a population with a non-constant variance. For instance, in the case where the relationship between price and size is more variable for larger properties, regardless of submarket location. Traditionally, this has been tackled by transformation, such as in weighted least squares regression (e.g. Harrison & Rubinfeld (1978)), but this is not modelling the heteroscedasticity directly within the specification. Multi-level modelling can overcome this by expanding the random part of the micro-model at level 1 (houses), to give a multi-level model with an additional random term for property size (Jones & Bullen, 1994):

$$P_{ij} = \alpha_j X_{ij} + \beta_0 Z_{0ij} + \Sigma \, \beta_k Z_{kij} + \left(\mu_{\alpha j} X_{ij} + \mu_{\beta k j} Z_{kij} + \varepsilon_{\beta 0ij} Z_{0ij} + \varepsilon_{ij} X_{ij} \right) \qquad 2.29$$

Hence, the attribute for property size, Z_0, is also included in the random part of the model.

Other features of the multi-level hedonic model Since the random-part differentials can be seen as coming from a distribution, it has certain practical benefits that are lacking in the traditional and spatial expansion specification estimated by OLS regression. Firstly, unlike the spatial expansion specification, which in effect fits a different regression model for each submarket, the multi-level specification uses all the data in the estimation. This means that the implicit prices of the attributes and the intercept term are based on information from all the submarkets. Secondly, this pooling of information results in 'borrowing of strength', whereby submarket-specific relations which are poorly estimated on their own benefit from information from other submarkets. Thus thirdly, this will result in precision-weighted estimation, whereby unreliable submarket specific implicit prices are differentially shrunk towards the overall market-wide estimate. A reliably estimated within submarket relation will not be affected by this shrinkage. Hence, these three features yield substantial benefits in implicit price estimation. Moreover, the pooling of information and borrowing of strength is more coterminous with the definition of submarkets as being quasi-independent and generating externality effects, more so than the switching regression or spatial expansion specifications, that in effect treat them as separate and autonomous. Hence, the estimation procedure is both specific and general, allowing an appropriate compromise between specific estimates for different places and the overall fixed estimate that pools information across places over the entire sample.

Extending the multi-level hedonic specification The multi-level specification may be elaborated upon in several ways. Firstly, the number of levels may be expanded to include different housing markets or different time periods. Secondly, additional attributes can be included in the micro-model, such as structural attributes, or in higher level macro-models, such as locational attributes. For instance, with respect to Can's (1990; 1992) work , if house prices are perceived to drift with neighbourhood quality, then this quality measure can be specified at the higher, submarket level. Hence:

$$\alpha_j = \alpha + \gamma \, NQ_j + \mu_{\alpha j} \qquad\qquad 2.30$$

In this random intercepts expansion, submarket average prices vary with neighbourhood quality.

A comparison of the contextual hedonic models Jones & Bullen (1994) compared the spatially expanded hedonic specifications with the multi-level specifications using house price data for London, divided into several districts (submarkets). A comparison of the estimates of the implicit attribute prices revealed that the multi-level specification produced more robust results. In particular, the multi-level model estimated coefficients gained from precision weighted shrinkage in submarkets that had small sample sizes, compared to the spatially expanded model coefficients estimated using OLS which 'does not differentiate whether two or two thousand houses are involved' (pp. 261). In this case, the coefficients had unreasonably high estimates. Moreover, several slope coefficients for house size in the spatially expanded specification were negative (suggesting paying less for a larger property), and these were shrunk to positive multi-level estimates in the majority of cases. Jones & Bullen (1994) demonstrated that the shrinkage was controlled by both the submarket sample size and the distance of the fixed-part estimate from the overall multi-level (London) average. It was also demonstrated that, unlike the spatially expanded counterpart, the multi-level specification was better able to deal with heteroscedasticity and autocorrelation by modelling it directly, and it was concluded that individual prices were indeed more variable for larger properties. Such variation could not be captured in the single parameter error term of the spatially expanded specification.

Conclusion

This chapter has reviewed the economic theory underpinning hedonic house price models, and has put the spatial element of such models into context. Specifically, it has discussed the problems encountered when using spatial data. Such problems have been generally ignored, and most recent research has concentrated upon using the parameter estimates as the basis for demand equations in an attempt to estimate the demand for housing attributes. However, if the hedonic model suffers from spatial effects, the estimated parameters may be incorrect and therefore so will the subsequent demand equations. To ameliorate these problems, alternative 'contextual' specifications were developed with the aim of modelling the spatial structures in the data explicitly. These were developed by expanding the fixed and random terms of the traditional hedonic specification to create the spatially expanded specification and the multi-level specification respectively. Although models using these specifications were advantageous over the traditional specification, it was concluded that the multi-level specification was conceptually and empirically more proficient when it came to modelling the spatial structures of the housing market. The next chapter will continue the theme of spatial data by examining the variable specification of the hedonic model. In particular, it will examine the problems associated with the measurement of locational attributes, and how Geographical Information Systems can overcome some of the issues of spatial resolution.

3 Housing Attributes and Spatial Data

Introduction

This chapter is concerned with the variable specification of the hedonic model, and some of the problems encountered with modelling such data. In chapter two, it was argued that housing attributes could be divided into structural attributes that pertained to the physical qualities of a house and locational attributes that are related to its location. The price of a property is then a realisation of the price of these attributes, although they are never individually identified in property transactions. However, it will be argued in this chapter that the measurement of these variables often fails to take into account the complexity of effects influencing house price, especially with respect to locational attributes, and underestimates the problems associated with the measurement of spatial data in general. The chapter is divided into four main sections. The first section deals with the types of attributes that affect house prices, with particular attention towards the concept of locational externalities. The second section deals with the measurement of housing attributes, within the context of previous research, with the emphasis upon the inconsistencies with regards locational attributes. The third section discusses the problems encountered with modelling spatial data using the hedonic price function, and how this has been historically limited by the resolution of the data. Finally, the last section discusses how Geographic Information Systems may overcome some of the problems associated with the measurement of locational externalities and the use of spatial data in general.

Housing Attributes

Introduction

Housing attributes have been traditionally divided into structural attributes and locational attributes. However, as will be seen, structural attributes have been far easier to account for in the price of the house than locational ones. Hence, a significant part of this section is devoted to locational attributes since, as will become apparent, they are far more difficult to reconcile than their structural counterparts.

Structural Attributes

Structural attributes describe the physical structure of a property and the land parcel within which it is located. Intrinsically, these attributes represent the shelter afforded by housing and the physical investment by the owner. Thus it can be argued that they provide the greatest utility to the consumer, and hence have the greatest weight in a utility function (Bajic, 1984). Furthermore, compared to locational attributes, structural attributes are conceptually more tangible and can be more accurately perceived. For instance, the number of rooms in a property is far easier to measure than is street quality. Structural attributes can be categorised into two groups (Follain & Jimenez, 1985. pp. 95. Table 2). The first contains attributes that pertain to living space; the second to structural quality. Although attributes of living space have commanded much more attention in the literature, Kain & Quigley (1970a) have suggested the structural quality of the housing bundle has at least as much affect upon price, and that 'understanding the nature of the urban housing market requires a better grasp of these kinds of interrelationships' (pp. 545).

Locational Attributes

> '[I]n the city, everything affects everything else' (Lowry, 1965. pp. 158 quoted in Harvey, 1973. pp. 58).

Introduction Locational attributes are measures of locational externalities. These have been defined as unpriced effects that affect the utility of others (Pinch, 1985. pp 8-9). They are unpriced in the sense that they are not paid for directly, but indirectly through housing purchase. They tend to be spatially concentrated in their impact upon the quality of people's lives and the value of their property. For this reason, externalities have been traditionally couched in terms of conflict between landuses that generate them and the residential location of different groups of people who may be affected by them (Cox, 1979). Pinch (1985) distinguishes between externality effects depending upon whether or not they impose a cost or benefit to the householder. These shall be discussed in turn.

Negative externalities A negative externality is an effect that creates disutility to the consumer. For instance, a factory may pollute the air in a residential neighbourhood and cause disutility to the households of that neighbourhood through poor air quality, ill health, and cleaning costs for which the factory does not offer any compensation. In other words, the factory externalises the costs of production (Cox, 1979). Amenities that are associated with such costs are often termed noxious facilities and tend to have a negative affect upon house prices.

Positive externalities These are the opposite of negative externalities. They are benefits received as a by-product of an activity to which the beneficiaries do not provide any direct payment. A good example of this is open space in an urban area which, although benefits all the residents of the urban area, has an additional benefit for the residents directly adjacent to it. As such, they tend to have a positive influence upon house prices. In some cases, a single amenity may emit both positive and negative externalities, such as a local shopping centre. This imposes negative externalities in the form of traffic congestion and noise pollution and positive externalities through convenience to the local residents.

Asymmetrical and reciprocal externalities It is also possible to differentiate between asymmetrical and reciprocal externalities (Cox, 1979). With an asymmetrical externality, the producer and consumer of the effect can be distinguished. In the case of negative externalities, the producers impose disutility and gains at the consumers' expense, such as air pollution from a factory. With positive externalities, the producer generates utility and the transfer of gain is in the opposite direction, as with public parks. In reciprocal relationships, the producers and consumers of externalities are the same. Traffic congestion, for example, is a negative externality produced by drivers and consumed by fellow drivers. They both produce costs for others and experience the costs imposed by others. In the case of positive reciprocal externalities, both producers and consumers benefit from each other. For instance, the residents in a neighbourhood may provide benefits for each other in the form of mutual assistance such as neighbourhood watch schemes. However, this asymmetrical and reciprocal dichotomy is rare, since most externalities are a mixture of both relationships.

The effects of externalities Most externalities are local in their impact, with a distance decay effect in their extent and intensity. Generally, households closest to the source of the externality will be affected the most, with the intensity of this effect diminishing with distance. A park will be of the most benefit to those households immediately adjacent to it, with this benefit diminishing rapidly with distance. Also, the larger the facility, the greater the intensity and range of these effects. Hence, large parks will have a greater and geographically wider influence on house prices than small parks. A major problem is how to measure the range of these effects. Many studies have used arbitrary thresholds around a facility to represent a catchment area, but this does not take into account the distance decay effects which will vary with the amenity (Pinch, 1985). In many cases the decay effect may not be a monotonic function of distance, since many externalities have both a positive and negative impact. For example, although a location next to a school may be beneficial with respect to accessibility, this benefit may be negated due to school related externalities such as noise and traffic. Therefore the distance decay function will be non-monotonic with the optimal location viewed as a trade off between the benefits of increased accessibility and the costs of proximity (Harvey, 1973).

Another problem is that the extent to which an externality is perceived as a cost or benefit will vary between individuals, and whether they use the facility or not. Air pollution from a factory will probably be perceived as less of a cost to those individuals who work at the factory than those individuals who do not. This implies that externalities will be positive for some members of society and negative for others. In attempt to clarify this conflict, Dear (1976) defined 'user-associated' and 'neighbourhood-associated' externalities. Generally, 'user-associated' externalities benefits the consumer who may live beyond the neighbourhood within which it is located, but not the non-users who live in the neighbourhood itself, and vice versa. In addition, Knox (1995) distinguishes between 'public behaviour' neighbourhood externalities and 'status' neighbourhood externalities. The former concerns the effect of peoples behaviour in public, such as tidiness, quiet and sobriety, whilst the latter relates to the reflected glory of living in a distinctive neighbourhood. Hence, the affects of location on price will also depend upon the income of different groups, their demography and their social attitudes.

Common Hedonic Model Attributes

Introduction

Although much is known about the attributes that affect the price of housing, theory offers little guidance in determining which of these to include in the hedonic model (Ohsfeldt, 1988). This is demonstrated in the literature, where the lack of widespread agreement has resulted in a diverse range of variables entering the hedonic specification. Furthermore, far more attention has been paid to date to structural attributes than to locational ones (Cheshire & Sheppard, 1995). All previous models have contained similar structural attributes to the extent that they represent the majority of the variables in the model. Graves et al, (1988) have categorised the variables included in the hedonic model in terms of being free, focused or doubtful. Free variables are those that are known to affect house prices, but are of no special interest in the study. These include certain structural and locational variables, such as floor area and distance to the city centre which may be of no particular interest to the researcher, but which will seriously bias the results if omitted. Focus variables are those variables of particular interest, and hence whose inclusion may vary from study to study. Doubtful variables may or may not affect the independent variable, but whose *a priori* omission may bias the results. Since the term doubtful variables have generally been applied to locational attributes, their importance in previous models has been very inconsistent, to such an extent that hedonic research can be classified on the basis of the inclusion of locational attributes. Hence, the following is an overview of previous research based on such a classification. This classification is also substantive in the sense that the treatment of locational attributes reflects the availability of spatial data.

Dwelling Specific Models

These models, also called location insensitive models (Ball, 1973), treat housing as essentially aspatial and disregard their geographic location as an unimportant factor in determining price. This suggests that households benefit from the structural attributes of the property only, and gain no utility from its location. Although this is theoretically and conceptually unsupported, structural attributes have been given prominence, particularly in early research (e.g. Cubin, 1970; Kain & Quigley, 1970a; 1970b; Wilkinson 1971; 1973b). The types of structural attributes used have varied, and to some extent seem to have been determined by data availability.

Following Follain & Jimenez (1985), Table 3.1 is a summary of the most common structural attributes from previous research. Measures of living space have been reduced to lot size, floor area and the number of rooms, whilst structural quality is concerned with age, style and interior and exterior quality. Previous studies have argued that floor area and number of rooms give sufficient information about the size and structure of the dwelling (Lineman, 1980). Both attributes are relatively non-malleable

Table 3.1 A summary of commonly used structural attributes

Classification	Attributes
Living Space	Interior Attributes
	Total Interior Living Space
	Total Floor Area
	Number of Stories
	Number of Rooms
	Number of Bedrooms
	Number of Recreation Rooms
	Number of Bathrooms
	Basement
	Attic
	Exterior Attributes
	Lot Size
	Off Road Parking
	Number of Garages
Structural Quality	Index of Dwelling Quality
	Age of House
	Presence of Full Insulation
	Brick Exterior
	Style of House
	Presence of Fireplace
	Double Glazing
	Air Conditioning
	Full Central Heating

Source: Adapted from Follain & Jimenez (1985) pp. 134.

and their supply is fixed in the short-run (Bajic, 1984). Also, to avoid overlap between lot size and size of the dwelling, outside lot area and internal floor area are often used instead of total lot size.

Structural quality is often harder to measure and is more subjective and corresponds to the physical condition of both the interior and exterior. The structural quality attribute is often constructed from several questions about the state of repair of the structure. For instance, Ohsfeldt (1988) constructed an index based on a scale of 0-6, determined by questions on state of repair and basic facilities. However, since an index of six simply measures a house in an average state of repair with basic amenities, this is actually an index of disrepair and deficiency as opposed to one that measures a range of structural quality . To rectify this, studies often include measures of enhanced quality such as central heating, double glazing and insulation to capture variations in structural quality of dwellings that do not have basic structural problems.

Wilkinson (1973a) and Ball (1973) may have been mistaken to describe these models as purely dwelling specific and locationally insensitive. Theoretically, a property's structural attributes and its location within the city are related, since they reflect the growth of the urban structure (Muth, 1969). This implies that an element of location will be inherent within the physical structure of the property. For example, although the age of the property, the size of building parcels and the patterns of tenure are all structural attributes, they will tend to vary systematically across urban space, reflecting historical growth patterns (Batty & Longley, 1987). Thus, such attributes may also proxy locational measures. This is indicated in studies such as Cubbin (1970) and Kain & Quigley (1970a), which revealed a high degree of multicollinearity between structural attributes, and the results suffered from spatial autocorrelation.

Location Specific Models

In most studies, locational externalities have been conceived in terms of relative and fixed locational attributes. Relative locational attributes are measures that reflect the externalities of the local neighbourhood and are unique to an individual property, such as street quality. Fixed locational attributes capture the location of a property with respect to the whole urban area, and pertain to some form of accessibility measure, typically access to the CBD. Taken together, relative and fixed locational attributes are said to equal the total utility of location (Krumm, 1980). Krumm argues that the important distinction between fixed and relative locational attributes is that a household may only change the level of the former through migration, but may affect the latter through the use of resources at that location. Hence relative locational attributes are produced, to some extent, by all households in the neighbourhood. Although this dichotomy has caused some disagreement with respect to which attributes fall into which category, it has generally been accepted and hence, Follain & Jemenez (1985) have classified locational attributes from previous studies into categories of (fixed) accessibility and (relative) neighbourhood quality measures.

Accessibility and Fixed Locational Attributes

Accessibility has been the most fiercely contested and the single most important measure of location in hedonic house price models. Traditionally, it has represented the measurement of the bid-rent curve as proposed in the micro-economics literature. It was this literature that was the stimulus behind much of the initial hedonic work, with Muth's (1969) pioneering work on estimating the rent gradient for Chicago being of particular significance (Dubin & Sung, 1987). Therefore it is not surprising that most of the hedonic research has been built upon the monocentric models of Alonso (1964), and later Evans (1973). These espouse the importance of the city centre as the major factor influencing land values, with the resulting bid-rent curve translated into a negative house price gradient. However, more recent work in this area has acknowledged the complexity of accessibility and residential location, and has suggested that the monocentric model be substituted for a multi-centric or polycentric model (Gordon et al, 1986). Such a model describes a city as having more than one, and usually several, population and employment centres, although the CBD is still envisaged to be the most prominent one (Griffith, 1981). Sanchez (1993) quotes changes in economic conditions, transportation costs, technology, and social patterns as giving rise to the sort of urban morphologies and land value distributions that are increasingly less 'center orientated' (pp. 455). It has often been claimed that these models are more representative of modern (US) cities, resulting in the hypothesis of an urban area with multiple house price gradients (e.g. Heikkia et al, 1989; Waddell et al, 1993). However, since much of the theoretical work is vague regarding the definition of non-CBD centres, the exact nature of what is being measured is questionable. For instance, although a polycentric model may depict an urban area as having several centres, each centre may have a different function, and as such, a varying degree of influence on the urban area (Griffith, 1981). This is in contrast with the monocentric city that has a single, well defined multi-functional focus that is hypothesized to have an affect across the entire city. This vagueness is also exacerbated by a lack of complementary empirical work, so that accessibility measures within a polycentric context have little to go on. These issues will now be expanded upon in the next section.

Measurement of Accessibility

The monocentric urban model The majority of hedonic house price research has been theoretically underpinned by this model, which proposes the existence of a negative house price gradient from the city centre reflecting the trade off between house price and declining accessibility. Cheshire & Sheppard (1995) argue that if locational attributes are appropriately measured, then monocentric models can perform well in a UK context. However this may not necessarily be true for modern US cities.

Various different measurements have been implemented to capture the affect of accessibility with equally differing results. Straight-line distance to the CBD has been the usual measure, with a linear or semi-log functional form, with a recent shift to route distance, although there is no strong argument that this is an improvement (Cooley et al, 1995). Physical distance is not the only measure of accessibility that has been used. Accessibility in terms of journey to work time, travel time and monetary loss have also been considered (e.g. Wabe, 1971; Bajic, 1984; Sanchez, 1993). In terms of the complex and numerous permutations reflected in the modal split found in most urban areas, measuring accessibility in terms of merely physical distance or one form of transport will undoubtedly oversimplify the problem. Also travel time may vary with hour of day and day of the week.

The functional form of the accessibility measure need not be smooth, or continuous since structural features such as arterial roads, rivers and railway lines can all distort accessibility. Moreover, transport systems, and so transport costs, are not necessarily the same across an urban area, so the accessibility function may vary with direction from the city centre (Cheshire & Sheppard, 1995). Accessibility measures such as these have been explored by Dubin & Sung (1987) and Waddell et al, (1993). They both divide the urban area into sectors and use dummy variables to allow shifts in the slope of the price gradient which would allow a complex functional form to be represented in a simple manner, and thus avoiding specifying the functional from in an *a priori* manner. Of course, the estimation of the functional form will depend in part on the dummy variables used. Waddell et al (1993) used arbitrary distance intervals, whilst Dubin & Sung (1987) hypothesized the location of breaks in the price gradient. Both conclude that these accessibility measures performed better than when standard functional forms, such as log of distance, were used.

Distance to transport routes, such as main roads, railways and bus routes have been an important feature of accessibility measures, since it is hypothesized that proximity will increase property values because of increases in accessibility and decreased transportation costs (Forrest et al, 1996). Of course, with respect to motorways and railways, it is distance to intersections and stations respectively that are of interest. On the other hand, transport systems generate negative externalities such as noise and air pollution and congestion which may eliminate any locational advantages of proximity (Sanchez, 1993). For instance, Waddell et al (1993) demonstrated that the highway proximity gradient in their study was non-linear, due to the associated negative externalities depressing property values close to a highway, but with the values increasing a short distance away once these externalities effects had become minimal.

However, previous empirical studies provide no consensus on the magnitude, or in some cases, the existence of a price gradient. It is quite common for empirical studies to fail to find a statistically significant between city centre access and price. Several studies have even reported statistically significant accessibility influences of the wrong sign, that is, price increasing with distance. These conflicting results are often attributed to the inadequacy of traditional accessibility measures (e.g. Goodman, 1979; Heikilla et al, 1989). In an overview of initial work, Ball (1973) particularly emphasized this lack of consensus, and he attributed it himself to poor data and the variety of measures used. However, he also suggested that there was no reason to assume that all cities should produce the same results, since the price gradient should be viewed within its historical and geographical context. For instance, Ball & Kirwin (1977) suggested the historical geography of the housing stock in Bristol, with affluent suburbs close to the city centre, resulted in the overall importance of accessibility in determining prices being smaller than anticipated. These contradictory results have since been blamed upon the theory of urban residential location that underpins it, and in particular the trade-off mode: 'It is not clear that the trade-off between commuting and land rent plays any significant role at all in location decisions' (Hamilton, 1982; pp. 1050).

The polycentric urban model Cities rarely have a simple monocentric structure, since employment and amenity centres are often located outside of the city centre, and this may cause the house price gradient to be complex, and undermine the significance of the city centre price gradient. Indeed, commentators such as Jackson (1979), Hamilton (1982) and Dubin & Sung (1987) have commented upon the implicit and often unjustified assumption of the monocentric city, and have criticised the negligible attention paid to the possibility of other influential sub-centres. Ball & Kirwan (1977) view the monocentric city as a 'major shortcoming' in a significant amount of hedonic research, whilst Hamilton (1982) asserts that the 'widespread acceptance of the class of monocentric models seem to rest more on its intrinsic plausibility than on any demonstrated ability to withstand empirical scrutiny' (pp. 1035). Furthermore, Bender & Hwang (1985) point out that the monocentric city is only a special case of the standard urban model. They argue that although the standard urban model proposes a negative house price gradient from the city centre, it also allows for the existence of secondary employment centres outside the CBD, which 'is not only intuitive but also empirically relevant' (pp. 91). Failure to control for commuting time to these secondary employment centres will tend to positively bias the coefficient of accessibility to the CBD. This, they suggest, probably explains the 'luke warm support' for a negative house price gradient in previous research (ibid). Hence, the urban area may be visualised as a price surface with a global maximum at the CBD and local maxima at the secondary employment centres (Jackson, 1979). This multiple-access hypothesis rests on the idea that the value of a property is determined in part by distance to each of these centres.

Empirical research on the nature of property prices within a polycentric urban context has so far been comparatively scarce, with the few exceptions including Jackson (1979), Bender & Hwang (1985), Dubin & Sung (1987), Heikkula et al (1989), and Waddell et al (1993). However, the definition of a secondary centre in such research has been vague. Typically, they refer to concentrations of non-CBD employment, but other urban amenity centres have also been considered. These have included shopping centres, hospitals, airports, and cultural centres such as universities. In fact, these features were hypothesized by Muth (1969) as the cause of local peaks in his estimated rent gradient. But it is difficult to determine how the traditional concept of accessibility, the measurement of the bid-rent curve, can be applied to some of these proposed secondary centres. The assumption is that, similar to the CBD, the secondary centres provide a large bundle of public services that are capitalised into property values. It can be hypothesized that this may be the case for employment centres, since these will usually generate a larger demand for labour in the local vicinity, relative to the CBD. Sanchez (1993) has further argued that, if the secondary employment centre is large enough, house price decrease will be affected by it, regardless of proximity to the CBD. However, it is questionable whether the other proposed secondary centres, such as airports and hospitals, will have the same effect. For instance, Waddell et al (1993) showed that the price effect of major secondary employment centres remained significant over a much larger area than more localised amenities such as airports and retail centres. Besides, as will be explained in the next section, some of these features have also be defined in terms of neighbourhood externality effects. Hence, there seems to be confusion in the literature between accessibility effects and the proximity effects of externalities.

Multiple accessibility measures There have been two main methods of estimating accessibility within a polycentric urban model. The first has measured the effects of multiple centres by estimating an accessibility trend surface (Jackson, 1979). Such a model holds housing attributes constant across an urban area, but allows the price of land to vary spatially as a result of demand for more accessible sites. The power of the polynomial of the trend surface represents the accessibility surface. For instance, a quadratic approximation represents a surface with a single maximum value for accessibility; the monocentric city. More complex surfaces with multiple local peaks can be represented by increasing the degree of the polynomial. Jackson advocated using a combination of R-square and F-tests to ascertain which degree of polynomial estimates the accessibility surface the best. The resulting surface can then be mapped. The advantage of such a procedure is that the secondary centres that have a significant influence upon house prices do not have to be specified *a priori*. In the case of Milwaukee, Jackson discovered that the price surface was centrally located with respect to manufacturing employment, and then decreased in all directions, but increased again towards secondary centres such as the university and peripheral manufacturing areas. The contour lines followed fairly regular concentric rings, with distortions caused by the transport system.

The second method is to identify secondary employment centres *a priori*, and estimate price gradients using traditional accessibility measures. The usually procedure was to experiment with various hypothesized secondary centres and select those that produced the optimal results based upon R-square values and statistical tests. However, Heikkilla (1988) argues that it will be intuitively expected for two or more distance measures to be collinear when confined to a plane, and hence multiple accessibility measures may be subject to problems such as multicollinearity. Also, there is no reason to expect every secondary centre to have an effect upon every house in an urban area (Jackson, 1979), and so the urban area was partitioned and each property allocated to its closet centre. The results of the multiple-access studies share the same diverse results as the earlier monocentric work. With the exception of Waddell et al, (1993) study of Dallas, a common result was the surprising failure of the CBD to exert a dominant influence on the overall house price gradient. In the case of Baltimore, Dubin & Sung (1987) concluded that 'the CBD appears to behave like the other [secondary] centres: it has an impact, but this effect is limited to a relatively small area' (pp. 204.) Similar results were concluded for Milwaukee (Jackson, 1979) and Chicago (Bender & Hwang, 1985) . In the case of the latter, it was concluded that 'in areas relatively close to a particular employment centre, accessibility to that centre is the dominant accessibility variable influencing price' (pp. 102), and that Chicago could be best characterised as consisting of a major monocentric city and two minor monocentric 'cities' with overlapping boundaries. Furthermore, Heikkila et al, (1989), in a study of Los Angeles, found that the CBD price gradient became positive and statistically insignificant once distance to multiple employment centres were explicitly included in the model. They concluded that the total lack of influence of the CBD was due to Los Angeles being a special case in terms of dispersed employment and population. This may be the case, but Waddell et al. (1993) pointed out that the study estimated a highly significant and positive coefficient for age of dwelling which was troubling since this is counter-intuitive. They suggest that this is indicative of collinearity between age and distance from the CBD, as discussed with dwelling specific models, and advocated the use of dummy variables.

However, the conclusions from these studies are very similar and suggest that maybe it is misleading to assume that accessibility to the CBD is of sole importance in all urban areas, particularly those of a large, dispersed character, and that it could be that the price of housing could be influenced by attributes of a more localised nature.

Neighbourhood Quality and Relative Locational Attributes

> 'Obviously the value of land in any city is not a function of distance from the city centre alone: there are other exogenous variable' (Evans, 1973; pp. 60).

Most studies have acknowledged that the neighbourhood within which the dwelling is located is an influential factor affecting house price. Indeed, Muth (1969) discusses several factors that affect house prices, other than structural attributes and distance from the city centre, and these in general can be described as measures of neighbourhood quality. Studies have tended to account for neighbourhood quality by either explicitly including neighbourhood attributes or by stratifying the sample using the neighbourhood as the basis. But, as with accessibility, there appears to be little agreement upon how neighbourhood quality should be measured and its inclusion seems to be based upon data availability. Hence, Ball (1974) comments that neighbourhood quality is used to cover a large, ill defined set of influences on house prices. Furthermore, it has been argued that neighbourhood quality as it stands is an unobservable attribute that can only be measured indirectly by the use of proxy measures (e.g. Davies, 1974). However this is debatable since many aspects of neighbourhood quality are tangible and have been quantified, such as in The English Housing Condition Survey (e.g. 1991). But it is correct to argue that much hedonic work have relied upon proxy measures instead of direct measures of neighbourhood quality. Graves et al (1988) described these measures as doubtful variables since being a proxy to a true measure, it is not known whether they affect house price.

Despite this, Dubin & Sung, (1990: pp. 98) have classified measures of neighbourhood quality used in previous research by three broad categories; measures of local public amenities, measures of the socio-economic status of the neighbourhood, and measures of neighbourhood racial composition. This classification illustrates well the broad measures used to capture neighbourhood quality, with the use of proxy measures emphasized by the lack of a category for explicit measures of environmental quality. In comparison, Mingche & Brown (1980) argue that many studies have included few, if any, 'location-specific' attributes and advocates measures of the 'micro-neighbourhood' that are defined in terms of aesthetic attributes, pollution levels and 'proximity', by which they mean accessibility to local amenities. These attributes can be regarded as direct measures of neighbourhood quality, and are more analogous to the definition of locational externalities. However, as will be argued in the next section, the social and racial composition of a neighbourhood may still be influential, in addition to the quality of the environment. As Wilkinson (1973a) summarised: 'Neighbourhoods can be measured in terms of the characteristics of the dwelling, the people, and the physical and social amenities which comprise them' (pp. 76).

Measurements of Neighbourhood Quality

Public amenities Attributes relating to public amenities are generally the most straightforward to measure and interpret since they are principally regarded as direct measures of neighbourhood quality and are easily quantified. Generally the better the quality of the service, the more highly valued it is and so is positively capitalised into house price. However, Dubin & Sung (1990) concluded that they discovered that services were relatively unimportant in contrast to socio-economic and racial composition of the neighbourhood. They admit that this finding was surprising in light of the emphasis on public provision in the literature. Most studies are invariably concerned with the quality of local schools, and hence common factors include the pupil/teacher ratio and average examination results (Fortney, 1996; Cheshire & Sheppard, 1995; Herrin & Kern, 1992). Other public amenities have included the amenities provided by public parks, (e.g. McLeod, 1984;), golf courses (Do & Grunditski, 1995), and the availability of local shops (Powe et al, 1995; Lineman, 1980). Also included in the public amenity category are measures of local property tax rates and local government jurisdictions, both of which can have a bearing upon the quality and cost of public service provision.

Socio-economic status Measures of the socio-economic status of a neighbourhood are less tangible, and have been classified as 'doubtful variables' (Powe et al, 1995). Typically measures have been constructed from census variables that relate to income levels, education, age and car ownership. Other indices include ACORN classifications in a UK context (e.g. Forrest et al, 1996; Collins & Evans, 1994) which are marketed as 'pen portraits' of the attributes of an area based upon the characteristics of the residents in them (Longley & Clarke, 1995). The crux of the uncertainty is whether the presence of high income households increase the value of certain neighbourhoods or whether certain neighbourhood features attract high income households (Sanchez, 1993). If the former is the case, then socio-economic status can be thought of in terms of a direct measure of neighbourhood quality. Members of high socio-economic groups are thought to be more desirable neighbours since they value the quality of the local environment greater than those in lower social groups and as a result may be prepared to make larger investments to maintain that quality. This is supported by Knox (1995), who suggests that environmental quality is closely tied to patterns and processes of investment and disinvestment and of social segregation. Alternatively, in the case of the latter, neighbourhood quality is regarded as being income elastic, so that it is likely to be given greater weight by higher income groups who are attracted by such attributes. In this case, socio-economic status represents a proxy for other attributes of neighbourhood quality such as low levels of air pollution, a low crime rate and high aesthetic surroundings. However, it is unlikely that socio-economic status will fall precisely into either category. It can be argued that high income households can afford to live in attractive residential areas, and if these areas are in short supply, will out-bid households in lower income groups who must settle for cheaper housing, often in less attractive surroundings. But at the same time, the clustering of income groups into specific areas will tend to create many of the neighbourhood features that each group finds favourable, and this will attract the same kinds of households. Therefore, it can be argued that socio-economic status is in fact a measure of 'public behaviour' and 'status' neighbourhood externalities, as described by Knox (1995), and should be included as a direct measure in addition to, and not to the exclusion of, other attributes of neighbourhood quality.

Racial composition The case for racial composition is even less clear. Again, there is a disagreement over whether race is a direct measure or proxy for neighbourhood quality. The case for a direct measure argues that discrimination against racial minorities reduces their access to housing and consequently causes the price of housing to be higher than in white neighbourhoods (Berry, 1976; Schnare, 1976). A similar argument follows, namely that people of the same race prefer to live in the same neighbourhood and hence houses in segregated neighbourhoods are more in demand than houses in integrated ones, and this will be capitalised into its price (Daniels, 1975). This has been called the 'taste for segregation' model (Bender & Hwang, 1985) but has encountered difficulty with respect to the definition of segregated neighbourhoods. If, on the other hand, racial minorities prefer white neighbours then the price of a house would increase as a monotone function of the percentage white (Waddell et al, 1993). However, if the racial composition of a neighbourhood simply reflects other characteristics such as socio-economic class, income and depressed surroundings then race is merely a proxy for neighbourhood quality. However, Dubin & Sung (1990) discovered that both race and socio-economic class were important, and that neither alone could sufficiently explain house price variation. The majority of studies into the influence of racial segregation on house price have been in a US context. It is arguable that in a UK context, racial composition may not be as significant a factor in the majority of urban areas due to greater social, cultural and economic heterogeneity within the majority of cities.

Environmental Quality

Introduction According to Richardson (1976), the quality of the environment is one of the most important determinants of a household's location, and thus the price they are prepared to pay for a property. Under the Mingche & Brown (1980) classification, environmental quality can be regarded in terms of the aesthetics of the local area, the pollution levels and also proximity to local amenities.

Aesthetic measures Environmental quality is often associated with open space, such as fields, parks and beaches. Although such features may have an amenity value, such as for leisure, a view of the feature may also be perceived as a benefit and hence will be capitalised into the price of a property. As Gillard (1981) argues: "Even when a park may not be used for recreation because of crime problems, it may still be valued for aesthetic reasons by residents with a view of the park" (pp. 217).

This can be thought of as the aspect of a property. Previous research into aspect has been focused upon features such as river views (Lansford & Jones, 1995; McLeod, 1984, Darling, 1973), forestry (Tyrvainen, 1997; Garrod & Willis, 1992) and shoreline (Brown & Pollakowski, 1977). McLeod (1984) discovered that river views were particularly important, and had a greater influence than a view of a park. These results were supported by Lansford & Jones (1995), Darling (1973) and Gillard (1981), although were disputed by Davies (1974) and Brown & Pollakowski (1977). Of course, it is not expected that results from different cities should be identical. As Gillard (1981) argued, the price is not dependent upon the intrinsic worth, but the supply relative to demand. In some urban areas, aesthetic views may be so abundant that they may be regarded as a free good. Other measures of aesthetic quality common in studies have included topography, with elevated areas being more desirable, and explicit measures of street quality such as condition of roads and pavements. A further measure includes population density, although this may be a proxy for attributes such as open space. Alternatively, certain views can have a negative effect upon property values. In particular, industrial, business and transportation land uses can have a negative effect upon property prices with respect to aesthetic qualities (Powe et al, 1995).

Pollution measures Pollution, and specifically air pollution, has been the focus of a great deal of hedonic research. Perhaps the most influential study is Harrison & Rubinfeld's (1978) study of air pollution in Boston caused by traffic and industry. Similar studies into air pollution have included Ridker & Henning, (1967), Lineman, (1980) and Palmquist, (1984). The other important pollution measure has been noise pollution, particularly with respect to airports (e.g. Levesque, 1994; Collins & Evans, 1994; McMillian et al., 1980; Goodman, 1979). In a summary of the evidence concerning the price effect of airport noise from a variety of studies, Nelson (1980) concluded that excessive noise depresses property values. Moreover, this effect appears to exhibit a considerable similarity in a variety of cities at different times. Other source of noise pollution, such as that caused by roads and railways, have also been investigated (e.g. Cheshire & Sheppard, 1995; Hughes & Sirmans, 1992; Krumm, 1980).

Proximity measures These correspond to measures of locational externalities which Mingche & Brown (1980) have argued have been lacking in most studies. The attributes in which proximity is regarded as being significant are access to public amenities such as schools and shops, non-residential activities such as industrial sites and open space. Since proximity can have both a positive and negative effect, they advocate the use of a non-monotonic distance function. This was corroborated by Waddell & Berry (1993), who concluded that a non-linear function was discovered for the amenities they measured, and can be expected where access is valued, but where immediate negative externalities over-ride the gain increased from proximity. The amenities included major and minor shopping centres, universities, hospitals and airports, and in accordance with externality theory, different shaped distance decay functions were estimated for each amenity, to reflect the fact that the effects of different amenities vary in their range and magnitude on property prices. For instance, Waddell & Berry (1993) discovered that in Dallas, major retail centres had only a slightly negative slope, and hence positive effect, in the first half mile, reflecting the capitalisation of access in price, but also the negative effects of immediate proximity. Once outside the influence of the negative externalities of the shopping centre, the slope increased in magnitude between two to five miles before decaying. Minor shopping centres had a similar distance decay function, but were smaller in magnitude and decayed more rapidly.

However, proximity has traditionally caused controversy within hedonic studies, due to conflicting results in both general and specific research into its effects. Hence, in accordance with theory, Do et al, (1994) found that neighbourhood churches had a significant negative effect upon house prices, with this impact decreasing with distance from the church. They attributed this to negative externalities associated with congestion and noise pollution. More generally, Kain & Quigley (1970a) found by factor analysis that the presence of commercial and industrial uses on a street had the expected negative effect upon property values. Similarly, Stull (1975), in a study of Boston, found that the median value of single-unit owner-occupied homes were negatively related to both the proportion of land devoted to multiple-family use and that devoted to industrial land use. Also, Stull found that, as the proportion of commercial land in a neighbourhood exceeds five percent, the values of single-family homes tend to fall. However, Grether & Miesszkowski (1980) discovered that the results of proximity in New Haven were mixed. Industrial and public housing areas had a significant, negative effect whilst minor commercial centres have relatively little effect. They concluded that non-residential land use *per se* has no systematic effect on housing values, and that landuse externalise may be very localised in their effect, so that they are a 'next-door' phenomena. A similar conclusion was reached by Li & Brown (1980), after distinguishing between the value of access to non-residential land uses from the externalities generated by their uses.

Conclusion

To conclude this section, it should be clear that locational attributes have been problematic in previous research. Firstly, the distinction between fixed and relative locational attributes is not clear, with amenities such as hospitals and airports being categorised in terms of both accessibility and local proximity measures. Secondly, it should be evident that neighbourhood quality is a tangible and quantifiable feature, although as previously discussed, in the majority of studies it has been inferred by measures of socio-economic status and racial composition.

Data in Hedonic Models

Problems with Data Measurement

The above discussion has highlighted several themes related to the data used in previous research. Firstly, it should be evident that there are marked differences between the scope and quality of measures of structural attributes when compared to locational ones. Whilst the potential set of structural attributes is limited and well researched, 'the set of available neighbourhood [attributes] is nearly infinite' (Cheshire & Sheppard, 1995; pp. 24). They argue that this is both because locational data are more difficult to collect and because it is less obvious, *a priori*, which locational attributes are relevant in determining prices. This has resulted in two problems. Firstly, Butler (1982) infers from this that any estimate of the hedonic price function will be misspecified to some extent because some of the relevant variables will inevitably be omitted, whilst Lineman (1980) comments that this missing variable bias is a particular problem for locational attributes. An omitted variable will violate the error term since it will capture the effects of the missing variable. In addition, this may also bias the included variables, since they may compensate for the omitted variable that can lead to erroneous interpretations of the parameters. For instance, although Sanchez (1993) used miles of street per square mile of land in each census tract as a measure of accessibility, he points out that this could also be a gross indicator of housing unit density and lot sizes in each tract.

Secondly, the problems of data availability have led to the use of surrogate or proxy measures. This has been illustrated with the case of socio-economic class, which has frequently been used as the only measure of neighbourhood quality. Also, such proxy measures are often constructed from aggregated data. In the majority of cases, the quality of the data collected for these aggregated spatial units are 'notoriously poor' (Anselin, 1988b; pp. 283), typically census data aggregated to wards or enumeration districts. The problems caused by poor neighbourhood quality attributes on measures of accessibility have been well documented, such as the research by Herrin & Kern (1992) which showed that better neighbourhood measures yielded significant improvements in the accessibility parameter. More importantly, the lack of good data measurement can cause standard econometrics to fail in numerous ways when applied in a spatial context. As discussed in chapter two, the presence of spatial aggregation, spatial externalities and spillover effects will separately, or in combination, affect the properties of the hedonic price function and statistical tests.

A general rule of thumb in the basic hedonic literature is that the more variables that increase the adjusted R-squared, the better. Thus, it is common to find more than twenty attributes in the hedonic price function although this can lead to severe problems with multicollinearity, especially if poorly specified data are used. Multicollinearity can be expected due to the inter-relationships between structural attributes and locational attributes (Ozanne & Malpezzi, 1985). Such was concluded by Powe et al (1995), who claimed that the high correlations between amenity variables, socio-economic variables and structural variables was the single greatest problems in their empirical research. Structural attributes will tend to be collinear because of the relationships between house size and house structure. Large houses will tend to have more bedrooms and recreation rooms for instance, than smaller houses. Inter-relationships between variables are more problematic for locational attributes. The relationships between socio-economic class and neighbourhood quality have already been discussed. Similarly, accessibility measures have been plagued by multicollinearity with neighbourhood quality variables. This factor has been blamed for poor and counter-intuitive results, which has resulted in the cries of 'what happened to the CBD-distance gradient?' (Heikkila et al, 1989). For instance, Goodman (1979) argues that negative externalities emanating from the CBD, such as noise and air pollution, will be negatively correlated with distance and this may cancel out the effect of accessibility. If 'the relationships are of similar magnitudes with opposite signs, it is possible for the distance term to be insignificantly different from zero and considered unimportant' (pp. 327). Moreover, multicollinearity between structural and locational attributes has long been a cause for concern. Kain & Quigley (1970a) appreciated that the quality of the neighbourhood is to some extent influenced by the dwelling stock, and commented that the difficulty in separating the two was 'perhaps the most vexing problem encountered in evaluating the several attribute bundles of residential services' (pp. 533). More recent examples includes the previously discussed critique of Heikkila et al (1989) work on Los Angeles by Waddell et al, (1993), who concluded that collinearity between age and distance variables had resulted in the counter-intuitive result for accessibility.

However, despite the fact that multicollinearity is a common problem of hedonic models, it is 'one which is often conveniently ignored' (Garrod & Willis, 1992a; pp. 65). With the few recent exceptions, such as

Powe et al (1995) and Forrest et al (1996), explicit tests for multicollinearity have been lacking. Powe et al (1995) advocated the use of the variance inflation factor (Pindyck & Rubinfeld, 1991), and omitted the variables that were shown to have high partial correlation coefficients. Forrest et al, (1996) used the Klein test that had been commended by Maddala (1992) as being the most rigorous means of assessing multicollinearity. It is more usual for either multicollinearity to be dismissed as unimportant, or the model is tested for robustness by omitting variables suspected of causing multicollinearity and the model re-estimated.

The Spatial Resolutions of Data

The distinction that most research has made between fixed and relative locational attributes may not be helpful. The concept of locational externalities blurs this distinction, and such was Minchge & Brown's (1980) argument when they advocated specific measures of the micro-neighbourhood. By doing this they supported a more precise specification of locational externalities; that is, one which captured the magnitude and distance decay nature of the externalities. This cannot be achieved with the use of poorly specified 'blanket measures', which are typical of the types of proxy data that have been used. However, it is not the case that improving solely the quality of the data will improve accuracy of the models. A problem also lies in the fact that the attributes are strongly related to each other, and improved data may not resolve this problem. A possible method to overcome this problem is to use the expansion and multi-level specifications that were described in chapter two. These specifications acknowledge that some of the attributes are inextricably bound together, and takes this into account when the data are modelled. Moreover, the concept of multi-level location infers that locational externalities will operate across different spatial scales. For instance, accessibility externalities affect a wider area than neighbourhood or street externalities.

This is an important concept, but one which appears to have been neglected in the literature. Instead, as was argued in chapter two, property prices have tended to be viewed as varying continuously in one dimension across urban space rather than over different spatial scales. Hence, it may be possible to overcome the data problems if the structure of property prices were to reflect the multi-level nature of urban space, and externalities were allowed to operate at different spatial resolutions. This was illustrated in chapter two at two levels: the property level and the submarket level. However, a street level may be inserted between these two, such that a hierarchy of three resolutions may be conceptualised, that intrinsically captures the externalities that operate within a local housing market. These are summarised in Table 3.2. The externalities in each level can be differentiated by how each are affected by the activities of

Table 3.2 A multi-level conceptualisation of locational externalities

Property Level Externalities	Accessibility to CBD
	Accessibility to Major non-CBD Centres
	Motorway Exits
	Railway Stations
	Shopping Centres
	Suburban Employment Centres
	Proximity Measures to Non-residential Landuses
	Parks
	Schools
	Industry
	Commercial
	Local Shops
	Recreational Centres
	Cultural / Educational Centres
Street Level Externalities	Street Environment
	Class of Street
	Street Quality
	Non-residential Activity
	School Catchment Areas
Neighbourhood Level Externalities	Housing Density
	Proportion of Non-residential Landuse
	Proportion of Open Space
	Quality of Local Amenities
	Social Composition
	Racial Composition
	Prestige / Desirability

households and the attributes of property, and also by the range and extent of their influence. Hence, since fixed locational attributes, such as accessibility to the CBD, and relative locational attributes, such as proximity to non-residential landuse, are unique for each property, they can be conceived as operating at the property level. However, locational attributes, such as street quality are influenced by the activities of residents in the street and hence can be regarded as a street level externality. This may also be the case for local amenities, such as parks, if the effect of their proximity is localised. Relative locational attributes at the neighbourhood level are those externalities are that effect prices across wider areas. For instance, the effects of racial and social composition of a neighbourhood may be seen as operating at this level, as well as more amorphous concepts such as desirability. An externality may also operate at more than one spatial scale. For instance, the externality effects of local amenities such as a school, may operate at the property level with respect to issues of proximity, and also at the street level with respect to the catchment area. Both these two effects will influence property prices. This is discussed in more detail in chapter six, whilst the concept of how structural attributes and locational externalities should be modelled is considered in more detail in the next section, which examines how Geographic Information Systems (GIS) can aid the estimation of hedonic house price functions.

GIS and Hedonic Models

'The true potential of Geographical Information Systems lies in their ability to analyse spatial data using the techniques of spatial analysis.' (Goodchild, 1988; pp. 76)

Introduction

It should be clear from the discussions in this and the previous chapter that one of the main problems encountered with hedonic house price models is the treatment of locational data, whether in terms of modelling of geographic space or in the measurement of locational attributes. These issues are representative of spatial data analysis in general (Anselin and Griffith, 1988) and can be linked to both the disregard of the potential problems posed by spatial data, and of the availability of well defined spatial data which has historically been poor. However, during the past decade, advances have been made which have gone some way to rectify these problems. Firstly, as was discussed in chapter two, the spatial nature of the hedonic house price function has been acknowledged in the new expansion and multi-level specifications. Secondly, there has been a marked improvement in the availability of spatial data at a much finer resolution than has previously been available. Thirdly, there has been the continued improvement of GIS, which can now store and manipulate much larger volumes of data at greater speeds and efficiency. The first issue of model specification has been dealt with in the previous chapter. The next chapter will discuss the types of data available for such analysis. Hence, the remainder of this chapter will discuss how GIS can improve the estimation of hedonic data.

GIS and Hedonic House Price Research

It would appear that GIS is an ideal medium to approach hedonic house price research for several reasons. It is capable of organising and managing large spatial datasets, such as those used in hedonic house price studies. Moreover, a GIS can handle these data at various spatial resolutions, such as at the level of the individual property and neighbourhood, which is important in the context of this research. A GIS also provides a valuable platform for spatial analysis, particularly with respect to the distance and proximity measures that have caused controversy in previous work. Finally, a GIS can aid the visualisation of the spatial data and map the results of the modelling. However, it would be easy to overstate the effectiveness of GIS in aiding such analysis, as has been done in the past, and hence some of these issues are considered in more detail below.

GIS and the Data Environment

Introduction Housing attributes represent a host of physical and socio-economic variables that are available at different resolutions. Hence, whilst structural attribute data are generally available for the individual property, it is more typical for locational attribute data to be aggregated at a higher level, such as census areas. A major defining attribute of a GIS is its ability to make such diverse data sets compatible (Flowerdew, 1991). A GIS can achieve this because it treats the attribute data of an object, and its location in geographic space, as two separate entities. By storing these two types of information separately in a database system, and allowing interaction between them, the data can be manipulated on the basis of either geographic location or attribute value. This flexibility is the power that underpins GIS. Furthermore, since a GIS is capable of storing these data at different scales, data can be geo-referenced and aggregated at various spatial levels. This is important if externalities are perceived as operating at different spatial scales, such as at the street level or the neighbourhood level. Thus, it would appear that a GIS is an ideal environment for storing and integrating the different types of data used in hedonic house price studies. But this situation is not as simple as this, since it is important to appreciate the nature of data within the GIS, particularly with respect that areal data.

Martin (1996) argues that because GIS is at least one step removed from reality, it is necessary to carefully consider the nature of geographic objects, as these are crucial to any subsequent use of GIS in answering geographical questions. He suggests that a model of the data environment should form the context within which any attempt to build a conceptual model of a GIS should sit. However, existing theoretical models of GIS have tended only to address the functions and component parts of the systems, without reference to the underlying data model, and its relationship to the real world. A fundamental distinction in areal data is that between data collected for artificial and natural areal units. Typically, physical data, such as area of parkland or length of street, are collected for well defined natural units. These units are natural since their boundaries are a meaningful spatial representation of the data in question. For instance, a street is a meaningful spatial object for data collected at street level. In contrast, socio-economic data tends to be collected for artificial areal units that may have very little relation to the underlying population. It has been widely acknowledged, for example, that census geographies often bear little resemblance to the underlying population they are attempting to represent. Hence, an important consideration of the data environment is that any spatial study heavily depends upon the 'nature and intrinsic meaningfulness' of the spatial objects studied, particularly with respect to socio-economic data (Martin, 1996. pp. 54). The scale of the analysis is very important in GIS, and holding an appropriate model of the geographic world is fundamental to any form of GIS-based analysis. Moreover, the variety of basic units have presented a major obstacle, especially with respect to data collect for artificial areal units. The patterns apparent in these data may be as much due to the nature of the collection units as to the underlying phenomena, and there is no direct way of comparing data collected for differing sets of areal units (Flowerdew & Openshaw, 1987). This is known more generally as the modifiable areal unit problem.

The modifiable areal unit problem This is a key problem in the manipulation of spatial data, and it is actually two distinct but closely related problems (Openshaw, 1984). There is the scale problem that is related to the level of aggregation of the data. There is also the aggregation (or zoning) problem, that is related to the fact that much of the data used in spatial analysis are based on areal units that do not have intrinsic meaning in relation to the underlying population, and hence are 'modifiable' and can be regrouped at any given spatial resolution. Wrigley (1995) summaries the scale problem as the tendency, within a system of modifiable areal units, for different statistical results to be obtained from the same set of data when that information is grouped at different levels of spatial resolution. The zoning problem relates to the variability in statistical results obtained within a set of areal units as a function of the various ways those units can be grouped at a given scale. In practice, the scale and zoning problems interact, whilst the zoning problem is also influenced by spatial autocorrelation.

Classically the problem has been observed in the magnitude of correlation coefficients between variables, which increases as the size of the areas involved in the analysis increases (e.g. Openshaw & Taylor, 1979). In terms of zoning, Openshaw & Taylor have demonstrated that, at any given scale, the zoning problem is likely to be sufficient to ensure that a wide range of statistical results are obtained, particularly with respect to correlations that have a tendency to change in magnitude and direction. More recently, the effects of the modifiable areal unit problem has been extended to include results of multivariate analysis, and it has been demonstrated that the goodness-of-fit statistic and parameter estimates are also sensitive to variations in scale and zoning systems (Fotheringham & Wong, 1991). In addition, an intrinsically related problem is the ecological fallacy. This arises when areal-unit data are the only source available to the researcher but the objects of study are individual-level characteristics and relationships. For example, when structural attributes for individual houses are only available as an average at the street level. In this case, relationships at a particular level of aggregation do not necessarily hold for the individual observations.

One solution to the modifiable areal unit problem, which has been implicit in much geographical work, is to assume that the problem does not exist. This was a common solution when spatial data were hard to come by, and there was little choice over the areal units. But many commentators have since argued (e.g. Martin, 1996) that in the context of GIS, this should not the case. The advent of GIS has increasingly allowed access to finer resolution data and in a digital form that accommodates the design of various zoning systems. Openshaw (1995) argues that this has led to a user modifiable areal unit problem (UMAUP), since there is now greater freedom to produce different zoning systems and hence a range of different results. Thus there is a need to design zones that are intrinsically related to the objects of study if the results are to be meaningful. For instance, within hedonic research it has been argued that it is important to capture neighbourhood submarkets as precisely as possible, since failure to do this can lead to structural instability and incorrect estimates. Previously, administrative geographies were used to represent submarkets, due to a lack of data at other resolutions. However, with GIS and finer resolution data, the ability to aggregate the data into specified submarkets is greatly enhanced, and thus, so is the ability to generate a whole range of hedonic prices based upon different zoning schema. Therefore, to be meaningful, the zoning system needs to be constructed such that it accurately reflects the submarkets and this may rely upon some underlying concept or theory - see chapter four. By doing this, Openshaw (1995) argues that the modifiable areal unit problem will disappear, since it is only a problem whilst the influence of zoning systems are ignored.

Spatial analysis and GIS The second major function of a GIS is spatial analysis, which Goodchild (1987) has defined as the statistical description or explanation of either locational or attribute information, or both. From this description, it would seem that a GIS would be an effective tool for estimating hedonic house price models. However, in recent years there have been many debates concerning the use of GIS in spatial analysis. Although this is still ongoing, it has been acknowledged that GIS and spatial analysis are inextricably linked (Gatrell, 1991), although exactly what role GIS has is still unclear (Openshaw, 1994b). This uncertainty is summed up by Rogerson & Fotheringham (1994):

> 'Although GIS may not be absolutely necessary for spatial analysis, it can facilitate such analysis and may even provide insights that would otherwise be missed. It is possible, for example, that the representation of spatial data and model results within a GIS could lead to an improved understanding both of the attributes being examined and of the procedures used to examine them.' (pp. 1-2)

This rather vague view of the role of GIS in spatial analysis can be accounted for by the fact that the main use of GIS has been one of data storage and management (Getis, 1994). Thus while GIS has been used extensively for storing, manipulating, transforming and visualising spatial data, its spatial analysis potential has been under utilised. Openshaw (1994a) argues that attention needs to be moved away from the view of GIS as solely one of Geographic Information Handling to one of Geographic Information Using. He continues this argument by providing ten basic rules for identifying GISable (sic) spatial analysis technologies. Three main spatial modelling issues are discussed within this framework. Firstly, the problems previously discussed of using data aggregated at various geographical scales, particularly those resulting from the modifiable areal unit problem. Secondly, he argues that the real secret of spatial analysis in GIS is that it should be able to handle special features of spatial information, rather than ignoring them. By 'special features', he is referring to the spatially structured nature of data precision and errors, since spatial data are rarely spatially random. Finally, he advocates GIS in assisting in exploratory data analysis, since he argues that users rarely know what patterns or relationships exist in the data, and that we are blind to the spatial patterns and processes that exits. In this way, the GIS can 'let the data speak for themselves' and suggest, with a minimum of pre-conditioning, what patterns might exist. (Openshaw, 1994b). The latter two aspects of spatial modelling shall be examined in more detail with respect to hedonic house price research.

GIS spatial analysis facilities There is a fairly substantial body of literature which concerns the interface between GIS and spatial analysis. These can be roughly spilt into two groups. Firstly, those concerned with the spatial summarization of the data, that is, the basic functions for the selective retrieval of spatial information within defined areas of interest, and the calculation of various summary statistics of this information. It is widely acknowledged that existing GIS offer a powerful array of techniques for spatial summarization, such as query facilities, Boolean operators, point-in-polygon and polygon overlay analysis, and various buffering techniques. Used alone, or in conjunction, these can identify and isolate specific geographic areas of interest and provide any relevant data concerned with the specified area. For instance, neighbourhoods with similar socio-economic and demographic profiles can be selected easily, whilst operations such as POINT-IN-POLYGON analysis can derive spatial information, such as average housing type in each neighbourhood, from the raw data held at different resolutions. However, Bailey (1994) asserts that such operations do not actually constitute 'spatial analysis', since they do not involve the analysis of patterns in spatial data, or the study of possible relationships between patterns and other attributes or features within the study region, or the modelling of such relationships for the purpose of understanding or prediction.

Much less has been written about the GIS operations with respect to such a definition of spatial analysis. Currently few GIS packages offer any capability of spatial modelling, in a statistical sense, of either raw or derived spatial data. Openshaw (1991; pp. 389) observes wryly that '[a] good GIS will today probably contain over 1000 commands ... but .. none will be concerned with what would correctly be termed spatial analysis rather than data manipulation'. Gatrell (1994) comments that although this is a slight exaggeration, the essence of his argument is correct. Nevertheless, spatial analysis 'tool-kits' have been incorporated into GIS over the past few years. For instance, in ARC / INFO, techniques such as network analysis, routing, location/allocation modelling, and grid-based analysis has become standard spatial analysis tools in recent years. However, in terms of statistical spatial analysis, GIS is still very much in its infancy (Bailey, 1994).

Instead, most attention has been focused upon the links between GIS and spatial analysis. Goodchild (1991) has characterised the general types of links between GIS and spatial analysis in terms of 'fully integrated', 'tightly coupled' and 'loosely coupled'. A fully integrated linkage occurs when statistical spatial analysis software is incorporated into the GIS package, although this has been very limited and surprisingly slow. More attention has been focused upon 'loose coupling' and 'close coupling' of GIS and statistical software. In the former, data are exported and imported seemlessly into and from the GIS and statistical package whilst with the latter, the GIS allows custom written statistical functions to be embedded within the GIS. 'Loose coupling' GIS have been the most common form of linkage, with the coupling to an external statistical package or graphics software usually achieved through ASCII files exported from GIS.

It should be clear that although GIS is a valuable tool in spatial analysis, it has been under utilised. This has been mainly due to the neglect of spatial analysis in general, and statistical spatial analysis in particular, since both are in many ways fundamental to the effective use and exploitation of GIS in may different applied contexts (Openshaw, 1991). In chapter two, the problems of spatial structures inherent in spatial data, such as spatial dependence and spatial heterogeneity were discussed with respect to hedonic models. In this case, it is not only important to know the attributes of a house in a particular location, but also the relationship of these attributes to the attributes of houses in other locations. Getis (1994) calls this proximal space and has connections with spatial dependence. Since this spatial aspect of the data can be handled conveniently in a GIS, Sinton (1992) identifies this as the chief area for GIS research in the future: 'I believe that the spatial inter-dependence among geographic entities is the theoretical linchpin of the GIS industry' (pp. 2-3).

In this sense, Bailey (1994) argues that the potential benefits of GIS are largely in facilitating the construction of proximity matrices between locations, known as W matrices, which are a necessary input to many the autocorrelation methods. Martin (1996) also notes that the fact that GIS is able to encode both location and attributes makes possible the development of techniques which incorporate explicitly spatial concepts such as adjacency, contiguity and distance that are important measures with respect to W matrices. Furthermore, the potential exists using the GIS to derive more sophisticated relationship measures between areal units which account for physical barriers, such as rivers, and network structures, such as roads. Such a technique would be beneficial within hedonic house price research since it has already been noted in chapter two that such models suffer from spatial autocorrelation caused by adjacency effects.

GIS in previous hedonic research The above discussion has illustrated the potential benefits of using a GIS in hedonic house price research. However, although GIS technology has been generally available for well over a decade, its use within hedonic house price studies has been rare, with the few recent exceptions including Waddell & Berry (1993), Waddell et al, (1993), Sanchez (1993), Cooley et al, (1995), Kennedy, et al (1996) and Lake (1996). The main use of GIS in these studies has been mainly to calculate distance and proximity measures in terms of both physical distance and time using tool-boxes such as NETWORK in ARC / INFO. GIS has also been used to calculate lot-size from digitised boundary data (Waddell & Berry, 1993), and the mapping of the error terms to determine the existence of spatial autocorrelation (Waddell et al, 1993). Hence, it would appear that GIS has been under represented in this field of research.

Conclusion

The aim of this chapter was to assess and evaluate the types of housing attributes that enter the house price determination process. In doing so, it has highlighted the problems associated with locational attributes. There has been very little consensus to the types of locational attributes that influence house price, and empirical evidence is contradictory. In particular, evidence supporting the existence of a negative rent gradient form the CBD outwards has been conflicting, and this has cast doubts upon the validity of the assumptions underpinning the micro-economic theory of housing markets. It would appear that locational attributes suffer from both conceptual problems and measurement problems. With respect to the former, traditional classifications of locational attributes based upon the concepts of relative and fixed location have been found to be problematic, with many attributes falling into both classes. This is acerbated by the measurement problems, with many locational attributes being historically poorly specified. This has caused violation problems in the estimation of the hedonic price function, such as multicollinearity and spatial effects. Together, these factors have contributed to inconsistent results and a general uncertainty to the influence of location upon house prices.

As a means of ameliorating these problems, to approaches have been identified. Firstly, the concept of locational attributes as being either relative or fixed has been replaced by the concept of locational attributes operating across different spatial levels. Such a concept allows locational attributes to behave more like externalities, with their influence determined by both magnitude of the attribute and proximity to the property. The second approach is to use a GIS as a median for the study. This will allow the measurement of locational attributes to be improved, and will also aid in the analysis and visualisation of the results. Therefore, chapter four is concerned with the construction of a context-sensitive GIS, with a particular emphasis upon data integration and the generation of locational attribute data.

4　Constructing a Context-Sensitive Urban GIS

Introduction

Valuing the built environment is a non-trivial exercise. It requires good quality data of a high spatial resolution and econometric techniques that are able model housing market process. Much of the previous research has failed to capture the intricate spatial structures resulting from housing market processes and have thus produced a plethora of contradictory results that confuse urban economic theory. This suggests that it is necessary to investigate how the spatial dynamics of a housing market can be modelled by the hedonic house price function as a pre-requisite to valuing locational externalities. This represents the first part of the research, described in chapter five. By implication, this will be a large, macro-scale study, with the emphasis upon modelling variations in supply and demand mechanisms rather than locational attributes. The latter is the emphasis of chapter six, which will use the hedonic models developed in the chapter five to investigate the impact of locational externalities in a smaller, detailed study. Thus the purpose of this chapter is to outline the case study are and identify suitable datasets. Integral to this is a discussion of how a context-sensitive GIS can be developed to manage the data and generate accurate measures of locational externalities.

The Cardiff Housing Market

Introduction

An important consideration is the delimitation of the housing market. It has already been discussed that meaningful spatial analysis demands the right sort of data for the right sort of units, since results from the analysis of spatially heterogeneous processes can be influenced by the choice of areal units. However, the theory of how housing markets are structured is complex, and the examination of markets have frequently failed to address their definition, composition and structure (Adair et al, 1996). Most of the empirical work has shown a considerable degree of variation in the definition of housing market areas. The spatial definitions of markets have ranged from sub-city areas, such as specific suburbs, to whole cities, local labour market areas and standard regions, depending upon the purpose of study and the availability of data. Nevertheless, the choice of boundaries should be meaningful.

In this respect, the chosen area of study in this research is Cardiff. Cardiff was selected since it has been used in work concerning the changing geographies of revenue raising (Martin et al., (1992); Longley et al., (1994)), the results of which have informed this study. The definition of the Cardiff housing market is the urban area bounded by the jurisdiction of Cardiff City Council. This contains both the inner city of Cardiff and its suburban hinterland. The actual boundaries are well defined and do not intersect any continuous built up areas. As such, the housing market can be regarded as autonomous and self-contained.

Background History

Introduction Cardiff is the capital of Wales and was founded by the Romans, with Cardiff Castle in the centre of the city occupying the site of the original Roman fort. The Castle was added to in the nineteenth century, when the city was rapidly developing as a port, exporting coal from the South Wales valleys to the rest of the world. Its development slowed markedly in the aftermath of the First World War when this export trade slumped (Daunton, 1977). Growth during this period was very rapid, and as a result the inner city today still exhibits considerable homogeneity of built form. More recently this homogeneity has been disrupted to some degree by Local Authority built estates and scattered pockets of infill redevelopment. Cardiff City Council has vigorously pursued a range of urban renewal programmes over the last couple of decades, and it is the implementation of such policies that the City Council has defined the 'Inner Area' to approximate this predominately nineteenth century urban core. The northern boundary of the Inner Area is defined by a major dual carriageway trunk road that physically splits the Inner Area from the suburbs. The eastern and southern boundary is formed by Cardiff Bay, whilst its western side is defined by the River Ely and green belt land. Together, the Inner Area covers approximately twenty four square kilometres. Beyond the Inner Area, the 'Outer Area' contains Cardiff's

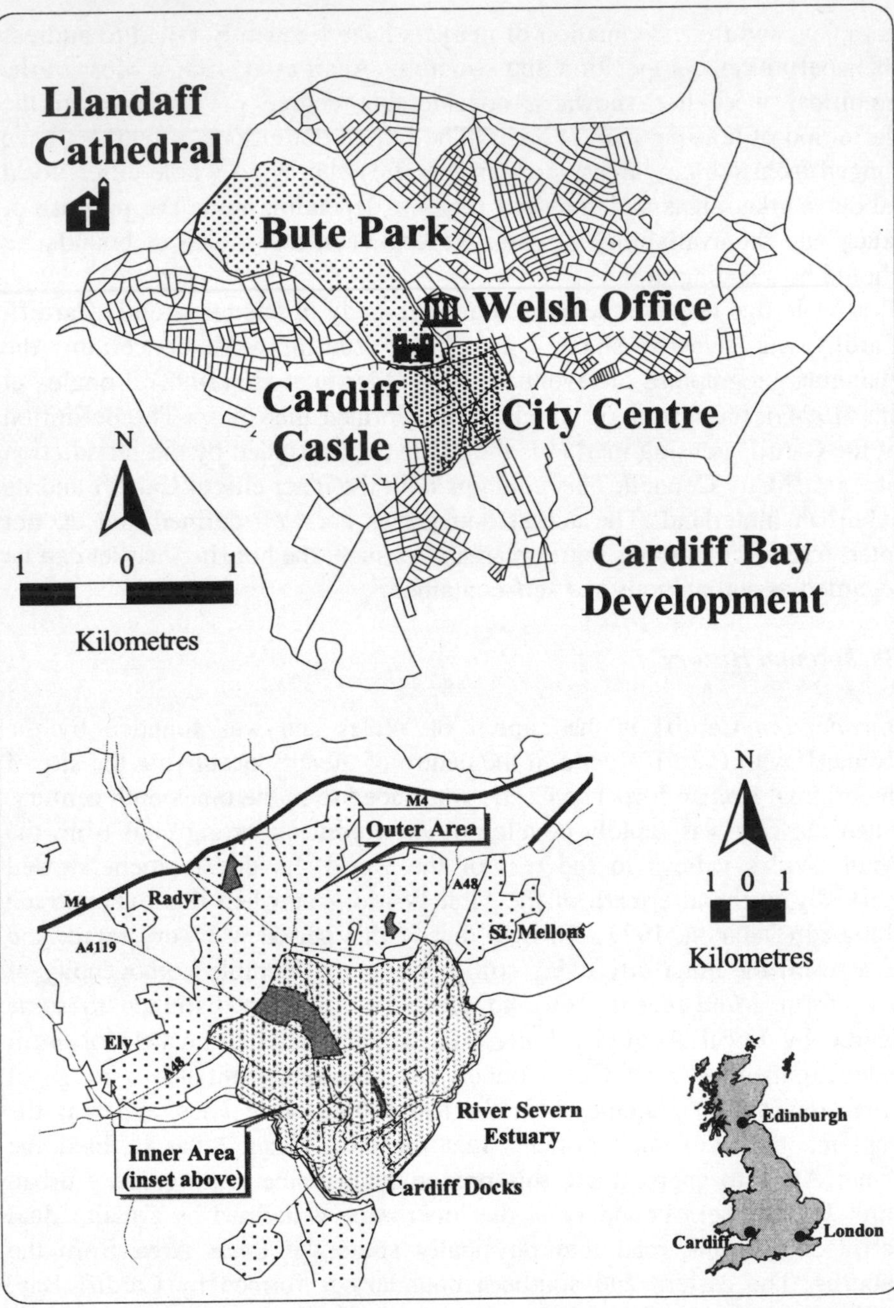

Figure 4.1 The built environment of Cardiff

suburbs - see Figure 4.1. There are considerable differences between the Inner and Outer Areas in terms of dwelling age and type, and differences in household size and composition.

The Inner Area The Inner Area is characterised by Victorian and Edwardian terraced housing, with early twentieth century semi-detached and detached private property located at its periphery. It has been identified by Cardiff City Council as a convenient unit for various aspects of urban policy, and remains a focus of sustained improvement and repair activity. There are a small number of Local Authority built housing estates, such as in Gabalfa and Tremorfa, and public sector purpose built flats in Butetown, although these do not represent a substantial proportion of the overall housing stock. The Inner Area also contains a small number of prestigious new housing projects, mainly associated with the Cardiff Bay redevelopment scheme. It is also the location of many older multi-occupied properties sub-divided into flats and bedsits. The Area includes the commercial and financial centre of the city, and is also the location of cultural and recreational amenities such as the Castle, Cathays Park, the City Hall, Law Courts, the Welsh Office, the National Museum of Wales, and the University College.

The Outer Area The suburbs contain lower density inter-war and post-war private housing and a number of extensive Local Authority estates. The western suburbs are characterised by the huge post-war Local Authority housing estates of Ely and Caerau, and also the prestigious neighbourhood of Llandaff. The eastern suburbs are the location of inter- and post-war semi-detached houses, with modern flats and apartments in Pentwyn and the sprawling modern peripheral estate of St. Mellons. The northern suburbs are represented by the upmarket neighbourhood of Cyncoed, containing the majority of Cardiff's bungalows, with modern housing estates beyond.

Cardiff's housing submarkets A second, related issue is the differentiation of the internal structure of the housing market. The principle of stratification of a housing market into subsets is widely recognised in the valuation literature (DeLisle, 1984) as a process of creating a number of homogeneous segments from a larger heterogeneous base. There are two broad approaches of stratification (Adair et al., 1996). The first is based upon the identification of distinct neighbourhoods that are readily distinguishable from one another primarily on the basis of environmental and locational characteristics. The second involves the identification of house groupings based upon differences in housing bundles, such as size, age and type. Both these approached were adopted, although it is only the former that is of interest in this section.

It has been previously noted that the operational definitions of submarkets have been problematic. Most researchers have used census or administrative geographies, primarily due to issues of data availability. To be meaningful, these sub-divisions have to relate to small scale supply and demand mechanisms, and hence property attributes and household characteristics. As such, the physical delimitation of census areas have often tried to capture existing neighbourhoods, particularly since census geographies often following defined boundaries such as street, rivers and railway lines. These can act as barriers to the movement of the population, and tend to fix the boundaries of local social interaction (Knox, 1996). With respect to Cardiff, 'communities' are the basic areal administrative units, and these are also electoral wards. The term 'community' suggests some form of internal social cohesion, and Cardiff estate agents often use them as a basis for defining residential neighbourhoods. Therefore it seemed sensible to use the same boundaries as an operational definition of submarkets for the whole of the housing market, resulting in twenty six submarkets in all - Figure 4.2.

Due to the obvious differences in housing stock and household composition, and its autonomous nature, it was decided that the Inner Area could be analysed separately, since the supply and demand schedules would be distinct from the rest of the city. The Inner Area is also an interesting area to study the effects of location since it is more heterogeneous than suburban locations, with property prices varying notably across smaller areas. It could be argued that locational externalities play a more significant part in price determination than in the Outer Area of Cardiff. The Inner Area contains nine complete communities, and the two partial communities of Roath and Llandaff. These in turn can be completely sub-divided into the eighty one Housing Condition Survey (HCS) Areas. These were defined by Cardiff City Council on the basis of within-area homogeneity of built form and residential characteristics in order to facilitate the detailed implementation of its housing policy.

Therefore, the housing market has been divided along two criteria. Firstly, the whole housing market has been stratified into twenty six submarkets based upon communities which local estate agents use as a rough definition of neighbourhood. Secondly, the housing market can be divided into an Inner and Outer Area, with the Inner Area corresponding to

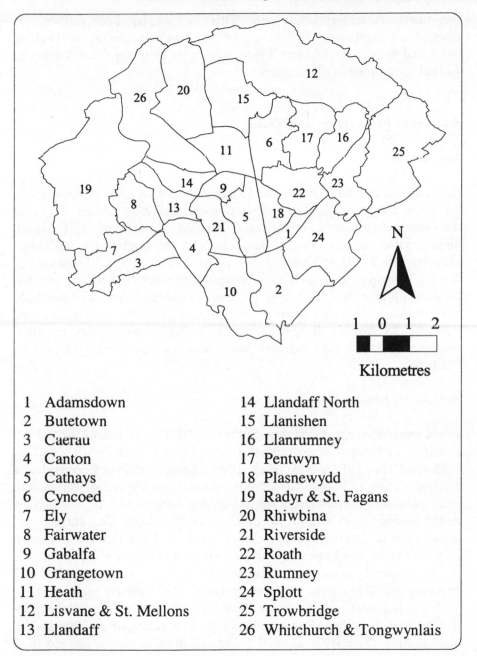

1	Adamsdown	14	Llandaff North
2	Butetown	15	Llanishen
3	Caerau	16	Llanrumney
4	Canton	17	Pentwyn
5	Cathays	18	Plasnewydd
6	Cyncoed	19	Radyr & St. Fagans
7	Ely	20	Rhiwbina
8	Fairwater	21	Riverside
9	Gabalfa	22	Roath
10	Grangetown	23	Rumney
11	Heath	24	Splott
12	Lisvane & St. Mellons	25	Trowbridge
13	Llandaff	26	Whitchurch & Tongwynlais

Figure 4.2 The location of Cardiff's communities

the nineteenth century urban core. This, in turn, has been further sub-divided into eighty one HCS Areas, based upon homogeneity of dwelling stock and resident population. These sub-divisions of the Cardiff housing market form the basis of the study.

Sources of Property Related Data

Introduction

Previous research has relied upon data that have been poor with respect to its spatial resolution, causing problems with hedonic modelling. To value the built environment effectively, detailed, locationally referenced, dissagregated data are needed, particularly at the level of the individual property. These data will have to incorporate both the physical structure of the built environment, such as structural attributes of housing and the location of amenities, and also its social structure, such as household characteristics. Although there is very often a difficulty in obtaining and managing locationally disaggregate data on eligible individuals and their residences, several data sources have been identified with respect to Cardiff.

The Cardiff Housing Condition Survey

The Cardiff Housing Condition Survey (CHCS) was commissioned by Cardiff City Council in November 1988, undertaken in the first half of 1989, and reported in November 1989 (Keltics, 1989). It provides a detailed picture of housing and environmental conditions, as well as the socio-economic characteristics of occupying households, of the private sector housing stock within the Inner Area of Cardiff. The survey was based upon an interval sample survey of 1 in 5 consecutive domestic dwellings within the Inner Area which were owned by either the occupiers, private landlords or housing associations. In total, a sample of 7,413 dwellings was drawn from a screened total of 37,115 private sector houses. Table 4.1 is a summary of some of the data in the CHCS. These have been divided into measures of street quality and measures of neighbourhood quality, since these were deemed to be the most useful to include in a hedonic model.

To aid the survey, the Inner Area was disaggregated into the eighty one housing condition survey (HCS) Areas. These aggregate precisely into communities, and the mean size of each area is approximately 450 dwellings. To date, it is the most detailed, locationally disaggregated private sector house condition survey ever carried out in British city.

Table 4.1 Summary of data in the CHCS

Physical Survey - Street Quality	Social Survey – HCS Area Quality
Front garden length	Quality of local shops
Condition of roads / pavements	Quality of public transport
Quality of street environment	Quality of access to city centre
Condition of rear lanes	Quality of sports facilities
Standard of upkeep	Quality of local parks
Traffic Problem	Quality of play spaces
Visually obstructive non-residential landuse	Quality of community facilities
Atmospheric obstructive non-residential landuse	Quality of physical environment
Noise obstructive non-residential landuse	
Derelict land / build in vicinity	

Source: Adapted from the Cardiff Housing Condition Survey, Keltics (1989).

Census Data

The 1991 OPCS census data were used to construct indices of deprivation and socio-economic classification at both the Enumeration District level and the Community (Ward) level for the whole of the Cardiff housing market. This follows similar methodologies used in many hedonic house price studies for constructing surrogates for locational attributes, and was necessary because data in the CHCS were only available for the Inner Area. The variables used to construct these indices are important, since they can determine the results of the analysis. The variables were chosen to represent the factors considered to influence housing supply and demand and residential differentiation (e.g. Hirschfield et al, 1995; Blake & Openshaw, 1996), and can be grouped into three general categories; socio-economic, family status/life-cycle and ethnicity; although it should be noted that these need not be mutually exclusive – see Table 4.2.

Table 4.2 Variables constructed from 1991 census data

Socio-economic dimension	Percentage of male unemployment
	Percentage of female unemployment
	Percentage of lone parent households
	Percentage of households with no car
	Percentage of households with two or more cars
	Percentage of households with no, or a shared bath or shower or inside WC
	Percentage of households with no central heating
	Percentage of households in owner occupied tenure
	Percentage of households in Local Authority tenure
Family life-cycle dimension	Percentage of households young and single
	Percentage of households pensioners
	Percentage of households married with family
Ethnic dimension	Percentage of non-white households

The Cardiff House Price Survey

Introduction Whilst the CHCS and census data provide information on locational attributes and residential composition, structural attribute data and property value data were collected for individual properties by means of a house price survey. The aim of the house price survey was to collect information on a cross-section of all properties in Cardiff, whilst ensuring that this cross-section was representative of all types of properties in all neighbourhoods. House price data were collected from local estate agents, whilst a dynamic sampling method was undertaken in an attempt to minimise any potential bias caused by under– or over-sampling of property types and neighbourhoods. Using this technique, around 1500 properties were sampled across the entire Cardiff housing market.

The estate agent data Table 4.3 summaries the property level information obtained from estate agency sources. The most important is house price. In most hedonic models, house price has usually been equated with selling price. However, within England and Wales a key concern relates to access to this information and this problem is two-fold (Estates Gazette, 1985). Firstly, due to legislative restrictions, information pertaining to sales and valuations of individual properties by financial institutions are confidential. Secondly, there is an absence of a central register of all sales. Only the Inland Revenue Valuation Office has full knowledge of all property dealings in England and Wales (Dixon, 1992), but this is not publicly available. Therefore, the asking price, which is freely available from estate agents, had to be collected instead.

Table 4.3 Data obtained in the house price survey

House Price	Full Central Heating
Postcode	Number of Garages
Total Floor Area (sq-ft)	Off-Road Parking
Dwelling Type	Age
Number of Bedrooms	Garden Size
Number of Recreation rooms	Swimming Pool
Number of Bathrooms	Conservatory
Number of Shower rooms	

In an attempt to be consistent with the data collected during the CHCS, the same categorisations were used for the structural attribute data where necessary. This included using the same categorisations for both property type and date of construction, whilst the same techniques employed in the CHCS were used for dating types of property and estimating garden size. The off-road parking variable represents any form of designated car parking space that was not covered by the garage category such as driveways, carports and specified parking bays. The presence of central heating was also recorded, as was whether the estate agent felt that the property was in need of substantial improvement.

The Cardiff Case Studies

The Cardiff Housing Market Macro-Scale Study

Introduction The aim of the first study is to model the spatial dynamics of the Cardiff housing market as a whole. It is hypothesized that supply and demand mechanisms will interact across the housing market at three spatial levels of resolution; the individual property level, the local neighbourhood level and the submarket level. To operationalise this, properties were geo-referenced using their postcode to a resolution of 100 metres. Traditionally, there has been concern about the accuracy of these grid-references since the system was not created with detailed geo-referencing in mind (Gatrell, 1989; Raper et al, 1992). A general consensus is that the grid references are far from optimal and in some cases are a cause for concern when used in small scale studies. However, for the purposes of the macro-scale study this should not be problematic. Similar to previous hedonic studies, the local neighbourhood level was proxied by Enumeration Districts, whilst submarkets were based upon communities.

The Inner Area Study

Introduction The Inner Area GIS is based around four spatial levels of resolution: the property level, the street level, the HCS Area level and the community level. These are seen as intrinsically capturing the externalities that operate at this local scale. The community level was added since these represent the neighbourhoods used by estate agents, and may be good proxies for submarkets.

The property level coverage In the macro-scale study each property is geo-referenced using its postcode. However, this is a very coarse method of geo-referencing property related data, particularly at small scales. Since the GIS needs to be as accurate as possible to capture local level locational externalities, a different method is used to geo-reference the property level data within the Inner Area. This method involves using the Ordnance Survey ADDRESS-POINT product as the basis of generating the property level coverage.

ADDRESS-POINT was launched by the OS in 1993, and provides a National Grid co-ordinate for each address at a resolution of 0.1m (Ordnance Survey, 1993). In addition, a unique Ordnance Survey ADDRESS-POINT Reference (OSAPR) code is given for each separate postal address. In some cases, separate postal addresses will share the same grid reference, but will have unique OSAPRs, such two dwellings on different floors of the same building. However, where there is only one delivery point in the building, and mail is sorted internally, there will be only one OSAPR relating to the one property. The OSAPR applies even when an address is changed, or the original structure is demolished and replaced. It only becomes 'dead' if the structure is demolished and not rebuilt. For the first time, ADDRESS-POINT represents a high resolution, fully comprehensive national property-level referencing system for the UK (Martin et al, 1994).

The street level coverage The street level coverage represents the second tier of the Inner Area GIS. A line coverage was generated by digitising the whole street network for the Inner Area of Cardiff (Martin et al, 1992). The coverage contains approximately 920 individual streets, and each was given a unique identification code based upon the street codes used in the CHCS. For the purposes of calculating accessibility measures, this line coverage was enlarged with the addition of main roads that connect the Inner Area to the M4 at the periphery of Cardiff.

The HCS Area and community level coverages Two polygon coverages were generated, representing the eighty one HCS Areas and eleven communities that comprise the Inner Area. Each polygon in the community coverage was assigned its ward code taken from the 1991 census and also its unique identification code in the CHCS. Each HCS Area polygon was assigned an identification code, also taken from the CHCS.

Constructing a Context-Sensitive GIS

Introduction An underlying problem in pervious hedonic research was the inability to model the built environment in sufficient detail. This can be resolved by using a GIS as a 'context-sensitive' means of handling complex urban datasets such as the CHCS and the house price survey data. Also, by inputting appropriate information, a GIS can generate measures of locational externalities, an important characteristic missing in the majority of studies. What follows is a detailed description of how a context-sensitive GIS was constructed for the Inner Area of Cardiff, with the aim of investigating the impact of externalities upon the value of housing within this area. The GIS was developed using ESRI's ARC / INFO GIS package.

Table 4.4 describes how the data is structured within the Inner Area. This indicates that the property level coverage will hold data pertaining to property valuations and structural attributes, as well as locational attributes relating to accessibility and proximity to non-residential landuses. The street level coverage will contain street environment data, and also data concerned with school catchment areas. The HCS Area level coverage will hold data relating to the local environment, particularly the quality of local amenities, whilst the community level coverage will contain data pertaining to social composition. Hence, the data from the CHCS, house price survey and census will have to be linked to the appropriate coverage.

Table 4.4 Housing attributes held in each coverage

Property Level Attributes	House price
	Structural attributes
	Accessibility measures to work place
	Proximity measures to non-residential landuses.
Street Level Attributes	Street environment measures
	Class of street
	Street quality
	Non-residential activity
	School catchment areas
HCS Area Level Attributes	Housing density
	Quality of local amenities
	Percentage of non-residential landuse
	Percentage of local Authority tenure
Community Level Attributes	Social composition

Linking address based datasets Since the CHCS and house price survey contain data collected at the level of the individual property, these need to be attached to the ADDRESS-POINT property level coverage. After investigating the address formats of each data set and ADDRESS-POINT, two procedures were considered:

1. the matching of ADDRESS-POINT and the data sets by text-strings;
2. the matching of ADDRESS-POINT and the data sets by a mixture of postcodes and addresses.

Matching text strings is particularly problematic given the variety of different address formats and conventions in existence in administrative data sets. Although the address format within ADDRESS-POINT correspond to the new British Standard BS7666 (Cushnie, 1994), this is not the case for the CHCS and house price survey data, where a variety of address formats and street and property naming standards were found to exist. Thus address based matching proved difficult. The second technique offers the possibility of isolating unit postcodes from both ADDRESS-POINT and the two datasets and, for each unit postcode, using the number of the dwelling as the 'match' since each unit postcode only contains approximately 12 - 14 addresses (Raper et al, 1992). This will allow properties that have differing text based information, such as '**37, Behets**

Avenue, CF2 7TY' and '37 Be-Hets Ave, CF2 7TY', and '16, Briony Street, CF2 5AD' and 'Flat 16 Briony St. CF2 5AD', to be matched. A computer programme was written that isolated the dwelling number and unit postcode of the addresses contained in all three datasets. Any postcodes missing in either the CHCS or house price dataset were subsequently added using the Postcode Address File. A second computer programme then matched these postcodes and dwelling numbers across the three datasets.

Results of the matching process This can be summarised by Raper et al's comment that 'the Royal Mail's task is certainly complicated by the fact that many properties ... have names, not numbers, and are amalgamations of former individual properties' (Raper, et al., 1992;. pp. 80). The matching process illustrated the complexity involved in matching address-based data as well as highlighting the types of errors likely to result from such matching. The most common errors relate to the postcode being recorded wrongly in the CHCS or house price data, or the address having a name or a composite number, such as 213-215 or 213/215 making a match on a single numeric problematic. Other problems occurred when the form of the address in the CHCS or house price data could not be related to the layout of the address in ADDRESS-POINT (e.g. **'Front Gnd Flr Flat 1 at No. 73 Diana Street'**) or where addresses contain alphanumeric descriptions (e.g. 47A). Those addresses in the CHCS and the house price data that failed to match by the computer programme had to be matched manually.

Some addresses that had been matched successfully still remained problematic. In the main, these were addresses associated with houses in multiple occupation and sub-divided properties, which only had one delivery point for the building and hence only one entry in ADDRESS-POINT, but more than one entry in the CHCS or house price data. In such a case, multiple matching between the datasets occurred. This also happened with blocks of flats that shared the same postcode. For example, since the numbers and postcodes of the properties in **'Flats 1-10 Stevens Court CF2 3RT'** and **'Flats 1-10 Thompson Towers CF2 3RT'** appear in the same combination twice, there will be multiple matching across the datasets. Of these various error types, manual checking resulted in considerable improvements, but this was very time consuming. Mismatches were spread throughout the Inner Area, but were particularly concentrated in those communities with a large proportion of sub-divided properties, such as Cathays. Recent work undertaken by the National Land

Information Service (NLIS) has highlighted similar problems when matching address based data sets to ADDRESS-POINT for properties in Bristol (Smith, 1996). Thus information loss appears to be a common occurrence with the integration of address related datasets.

Nesting the four GIS coverages To be able to model the spatial dynamics of the Inner Area housing market, the four spatial levels have to nest perfectly. Using ARC/INFO, each property in ADDRESS-POINT was located within a specific street, HCS Area and community. However, nesting the street coverage into the HCS Areas proved to be problematic, since many of the longer streets traversed the HCS Area boundaries. There was also the additional problem that the street network had been used as a template for the construction of HCS Area boundaries. This had resulted in many of the HCS Area boundaries running down the centre of a street, and hence the difficulty of allocating such a street into a specific polygon. The problem of nesting streets into HCS Areas was overcome by the construction of 'substreet' sections. These were the sections of street that fell wholly into each HCS Area. In most cases (73%), the street and substreets were equivalent, since they both fell within one HCS Area. However, in the other cases (27%), a single street was split into more than one substreet, either because it traversed a HCS Area boundary or because the boundary ran down the centre of the road. In the case of the latter, properties on opposite sides of the road were in placed into different substreets since they were in different HCS Areas. Since HCS Areas nested perfectly into communities, there was no difficulty of nesting substreets into communities. The resulting GIS is illustrated in Figure 4.3.

Generating substreet and HCS Area level data The CHCS contains data relating to the street environment although these are held at the property level. Using the GIS, these data were aggregated to the substreet level. It was decided to use substreet sections as the basis for the aggregation as opposed to actual street sections, since many of the longer streets in Cardiff will experience significant changes in street quality along their course. The street quality variables in Table 4.1 were used to create two street quality attributes; one measuring overall environmental quality, the other measuring the presence of non-residential landuse in the street. The first attribute was constructed by combining the first six street quality attributes in Table 4.1, to produce an overall index of street quality divided into four categories; poor, below average, above average and good. The second attribute was constructed by combining the remaining four street

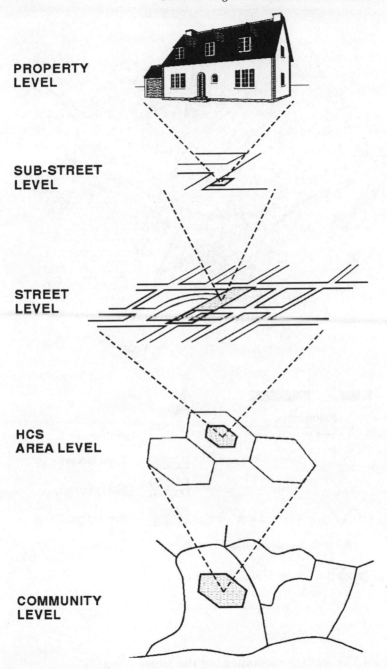

PROPERTY
LEVEL

SUB-STREET
LEVEL

STREET
LEVEL

HCS
AREA LEVEL

COMMUNITY
LEVEL

Figure 4.3 The nested Inner Area GIS

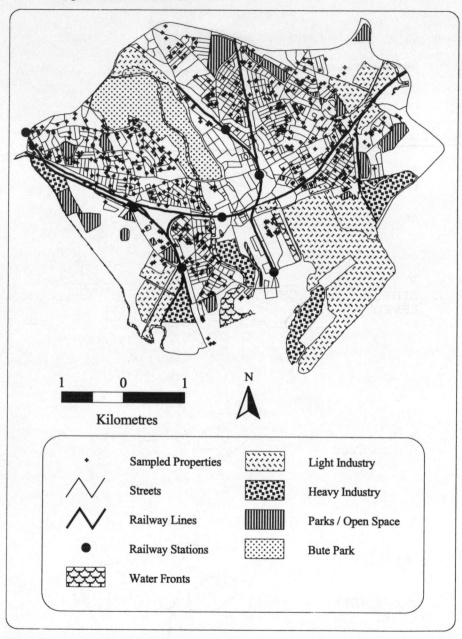

Figure 4.4 The built environment of the Inner Area

quality attributes in Table 4.1 to create an index that measured the presence and obtrusiveness of non-residential land use in the street ranging from 'not present' through to 'very obtrusive'. Using a similar method, data pertaining to local amenities in Table 4.1 were aggregated to the HCS Area level and linked to the HCS Area coverage. Finally, the census data at ward level were attached to the community coverage using its ward code.

Constructing Inner Area landuses Following pre-requisite fieldwork, the Inner Area's non-residential landuses were divided into three main categories: parks and open space, industrial areas and institutional landuses. The parks and open space category was further sub-divided to separate Bute Park, the city's principal open space and recreational area. In a similar manner, industrial landuses were sub-divided into 'heavy' and 'light' industrial areas. The former corresponds principally to the traditional manufacturing and extractive industries associated with the docks. The latter corresponds to modern trading estates that are generally devoid of these traditional types of industries. Institutional areas form the final major landuse category. These correspond to non-residential buildings and activities, such as the Government Offices and the University. Furthermore, Table 4.5 shows how these landuses have been further differentiated by their size. This will be taken into account when externalities are calculated.

Point coverages were constructed to capture the locations of other non-residential landuses such as schools and hospitals. In addition, information on pupil size and examination results were acquired for all the secondary schools in the Inner Area. Finally, the location of the railway lines and the River Taff were added to the Inner Area GIS. A point coverage representing the location of all the seven railway stations that serve the Inner Area, and a separate coverage consisting solely of Cardiff Central, Cardiff's principal railway station, were also added. Figure 4.4 shows the resulting composite GIS, illustrating how the built environment of the Inner Area can be abstracted.

Generating Locational Externalities

Introduction

One of the primary purposes of the GIS is to generate locational specific attribute data. Such data will allow the exploration and evaluation

Table 4.5 Classification of non-residential landuse in the Inner Area

'Light' Industry	Area (km²)	Park / Open space	Area (km²)
Gabalfa Industrial Estate	0.29	Butetown Open Space	0.03
Ipswich Road Industrial Estate	0.12	Channel View Playing Fields	0.05
Selan Industrial Estate	0.13	Coronation Park	0.01
Colchester/Dominions Way Industrial Estate	0.21	Grange Gardens	0.01
Clydesmuir Industrial Estate	0.10	Jubilee Recreation Ground	0.02
East Moors Trading Estate	0.82	Lansdowne Hospital Fields	0.11
Jubilee Trading Estate	0.04	Lawrenny Avenue Playing Fields	0.09
Portmanmoor Industrial Estate	0.41	Mill Gardens	0.01
Ocean Park trading Estate	1.65	Moira Place Gardens	0.01
Hadfield Road Trading Estate	0.13	Moorland Park	0.02
Leckwith Industrial Estate	0.20	Ninian Park Athletics Ground	0.03
'Heavy' Industry		Roath Park	0.11
Ace Industrial Estate	0.22	Sevenoaks Park	0.05
Butetown Works	0.25	Splott Park	0.06
Leckwith Industrial Estate	0.16	Thompson Park	0.05
Queen Alexandra Dock	0.39	Tremorfa Park	0.13
Seawall Road Industrial Estate	0.36	Victoria Park	0.06
Institutional		Waterloo Gardens	0.01
Crown Way Government Offices	0.31	Waterloo Recreation Ground	0.05
Welsh Offices	0.27		
Cardiff Institute of Higher Education	0.02		
Cardiff University	0.44		

of the two concepts of locational externalities: that their effects upon property values diminish with distance, and that they operate over different spatial scales. The construction of the locational attribute data is a two-fold process. Firstly, attribute data has to be generated for the whole of the Cardiff for use in the macro-scale study, although in effect, the use of the GIS in this process will be minimal. Instead, the GIS will be extensively used in generating detailed locational attribute data for the Inner Area housing market study.

The Cardiff Macro-Scale Study

Introduction Since the purpose of the Cardiff housing market macro-scale study is to explore spatial housing market dynamics within conventional urban theory, traditional measures of location were used. This involved the construction of a measure of accessibility to the city centre and measures of neighbourhood quality from census data at the Enumeration District level.

Access to the city centre A basic function of a GIS is to measure the Euclidean distance between two points. In the macro-scale study, accessibility is reduced to a simple measure of straight line distance from each property to Cardiff city centre. This was calculated within ARC / INFO using POINTDISTANCE. To ameliorate the problems of estimating the rent gradient when the functional form is not known, these linear accessibility measures were then divided into distance intervals. This has the advantage that the rent gradient is not constrained by any *a priori* imposed functional forms, whilst local maxima and minima may also be derived. Table 4.6 is a summary of the distance intervals. The choice of intervals are based upon the assumption that the rent gradient will be generally non-linear and decrease at a decreasing rate from the city centre. Hence, the curve will be at its steepest within the first couple of kilometres, after which it will begin to flatten out.

Neighbourhood quality measures Locational attributes quantifying neighbourhood quality were constructed from the census data in Table 4.2 Since a high degree of collinearity was present between the variables principal components analysis was used to construct new indices that would proxy locational attributes more effectively. In addition, the 'Percentage of households in Local Authority tenure' variable was omitted from the principal components analysis, and used to construct a separate measure to capture the 'stigma' of Local Authority built housing stock.

Table 4.6 Distance intervals from Cardiff city centre

Distance (metres)	Sample	Distance (metres)	Sample
0-100	6	2000-2200	27
100-200	7	2200-2400	27
200-300	11	2400-2600	14
300-400	10	2600-2800	24
400-500	27	2800-3000	26
500-600	33	3000-3200	44
600-700	43	3200-3400	25
700-800	35	3400-3600	41
800-900	25	3600-3800	23
900-1000	27	3800-4000	42
1000-1100	44	4000-4200	54
1100-1200	46	4200-4400	59
1200-1300	49	4400-4600	35
1300-1400	36	4600-4800	52
1400-1500	31	4800-5000	57
1500-1600	35	5000-5500	89
1600-1700	44	5500-6000	100
1700-1800	19	6000-6500	69
1800-1900	26	6500-7000	55
1900-2000	20	7000-9000	43

Principal components analysis This is a synthesizing technique that is able to identify groups of variables that have similar patterns of variation. Once identified, these groups of variables can be transformed into hybrid variables called components that summarise the original data. They are uncorrelated and each accounts for a successively smaller amount of the co-variance between the original variables. Hence, the technique allows variables that have a high degree of multicollinearity to be transformed into new, uncorrelated components. With the exception of the 'Percentage of households in Local Authority tenure', the variables in Table 4.2 underwent principal components analysis, the output of which is summarised in Table 4.7.

Table 4.7 Summary of the Principal Components Analysis

Eigenvalue	3.39	1.14	0.63	0.53	0.23	0.08
Proportion of explained variance	0.56	0.19	0.10	0.09	0.04	0.01
Cumulative explained variance	0.56	0.76	0.86	0.95	0.99	1.00

	Components		
Variables	1	2	3
Percentage male unemployment	-0.494	-0.144	0.056
Percentage female unemployment	-0.254	-0.051	0.032
Percentage households non-white	-0.312	0.509	0.655
Percentage households shared / no amenities	-0.509	0.728	-0.277
Percentage of households with no central heating	-0.556	0.486	-3.71
Percentage households owner occupied tenure	0.501	-0.098	-0.269
Percentage households no car	-0.509	0.104	-0.269
Percentage households two or more cars	0.468	0.104	0.440
Percentage households young/single	-0.402	0.517	0.176
Percentage households old	-0.338	0.445	-0.214
Percentage households families	0.214	0.179	-0.165
Percentage households lone parent family	-0.363	0.475	0.475

The first part of the table represents the eigenvalues of the correlation matrix, together with individual and cumulative percentages of the total variance. This suggests that three quarters of the variation within the census data lies in two-dimensional space, whilst 86 % lies in three-dimensional space. Higher dimensions each contain only a negligible proportion of the total variability. Since the objective is to identify only the major dimensions of co-variance in the data, it is usual to retain only those components that account for a greater proportion of the total variance than could any of the original variables. Following Mather (1976), it was decided that the first two principal components were satisfactory in describing the data on the grounds that they accounted for three quarters of the total variation, that they both had eigenvalues in excess of one, and the two components were interpretable with respect to the original data.

The second part of Table 4.7 summarises the loadings of each component. These represent correlations between the principal components and the original data, and indicate how much a variable has contributed to the construction of a particular component. In the first component, the range of the magnitudes of the loadings is quite small (0.214 - 0.556), suggesting that several variables have had an equal contribution. However, the highest loadings are associated with indicators of income. It is negatively correlated with areas of poor housing conditions, high male unemployment, single / young persons households, and a low degree of car ownership. Conversely, it is positively correlated with areas of owner occupation and two car households. The geography of component one indicates a concentration in rundown inner city neighbourhoods and peripheral council estates. Both these suggest that component one can be classified as a general measure of social economic class.

The loading in the second principal component can be allotted into three groups of significant variables. The highest loading identifies areas of housing in poor condition as being significant. Next are areas that have a high percentage of non-white households and young / single households, followed by areas that have a high percentage of elderly, and single parent households. It can be seen that unlike the former, these loadings are quite diverse. However, inspection of the data suggests that they can all be associated with areas of poor housing condition, the first loading directly so. This is qualified by a map of the component. Areas of strong, positive association are located in the peripheral council estates, and the inner city neighbourhoods to the east of the city centre. The CHCS concluded that these neighbourhoods contained a high concentration of houses in need of modernisation. Hence, although the explanation of component two is more vague compared to component one, it can be concluded that it is generally associated with areas of housing in a relatively poor condition.

Areas of Local Authority built stock The percentage of households in Local Authority tenure was used to construct a dummy variable representing EDs that have over 50 % of their housing stock in Local Authority ownership. This attribute reflects the stigma, suggested in literature and by estate agents, that affects areas of housing stock of Local Authority origin. The 50 % cut off was selected since analysis of the data suggested that this represented a meaningful break in the distribution. Also it was assumed that such stigma would only be perceptible in areas where houses of Local Authority origin dominated the housing stock.

The Cardiff Inner Area Study

Property level externalities Property level externalities include measures of accessibility and proximity to landuse externalities. With respect to the Inner Area, a more accurate measurement of accessibility is needed that takes into account the underlying typology of the road network. In such cases, accessibility between two points is the shortest route on the road network connecting them. An additional advantage of using the road network to measure accessibility is that the shortest route need not be measured solely in terms of distance. Other costs, called impedance costs, can be taken into consideration. Impedance costs take into account factors that may affect accessibility, such as speed limits. Impedance costs were assigned as estimated travel times along each street. Roads that have a lower travel time and thus impedance cost will be favoured when calculating accessibility.

The CHCS suggested that accessibility to Cardiff city centre, the M4 motorway and the rail network were all influential destinations for the majority of Inner Area residents. Accessibility to the city centre was calculated using the ALLOCATION command in the NETWORK module, working from the city centre outwards along the street network. This allowed the minimum travel time from the city centre to each property to be calculated simultaneously. Moreover, the ALLOCATION command allows more than one destination to be selected at any one time. Hence, accessibility to the nearest point of interest, such as the nearest railway station, can be calculated simultaneously for each property. Using this method, access to the CBD, the nearest M4 motorway junction and the nearest railway station in Cardiff were calculated for each property. A separate accessibility measure was also calculated for travel time to Cardiff Central railway station, the city's principal rail terminus.

The externality effects of non-residential landuses will depend upon their distance from the property and their relative attractiveness as an amenity. Both must be modelled simultaneously. This can be done in ARC / INFO by using the ACCESSIBILITY command, which calculates a measure of proximity as directly proportional to the supply of attributes at a given location, and inversely proportional to the distance away from it. This can be summarised thus:

$$P_i = \Sigma W_j \, d_{ij}^{-\beta} \qquad\qquad 4.1$$

where: P_i is the proximity of property i
W_j is the attractiveness of externality j
d_{ij} is the distance between property i and externality j
β is the exponent for distance decay
n is the number of externalities being measured

The attributes of the externality are used to compute an attractiveness index, whilst the effects of distance are scaled using a distance decay function β. The process of finding the value of β in computing ACCESSIBILITY is called calibration, and is a crucial part of measuring the externality effect. If the exponent is small, the effects of the externality increase and vice versa. Since the β value is not known *a priori*, a range of values were used to generate a range of externality measures for each landuse. These can then be calibrated within a hedonic model and the optimal exponent for each landuse estimated. The attractiveness index used in the computation was the area of land squared, except for those landuses represented by a point coverage, in which case the attractiveness index was set to unity.

Proximity measures to the River Taff and the railway lines were calculated in a slightly different way. By using previous studies as a yard stick (Lansford & Jones, 1995; McLeod, 1984), it was hypothesized that the externality effects associated with these features would be quite small. To quantify these, four sets of buffer zones were generated at intervals of fifty metres from each externality. The POINT-IN-POLYGON command was then used to determine which properties fell within fifty, one hundred, one hundred and fifty and two hundred metres of the river and railway lines respectively.

Street level externalities It is anticipated that the two measures of street quality generated from the CHCS will affect all properties located within the immediate substreet. However, since it is not known *a priori* how a locational externality diminishes with distance, the quality of one substreet may also influence properties in adjacent substreets, depending upon the perceptions of the buyer. To account for this, a range of street quality externality measures were calculated, using 'topologically sensitive' buffer zones of various sizes generated around each property (Orford, 1998). Although the size of each buffer zone would theoretically depend upon the extent of the influence of street quality on house prices, in practice the buffer zones were restricted to 50 metres, 100 metres and 200 metres.

Hence, each property has a measure of street quality based on two attributes at three spatial resolutions (0 - 50m, 50 - 100m and 100 - 200m). Secondary school catchment areas had to be approximated using thiessen polygons. Properties were placed into a catchment area using the POINT-IN-POLYGON command.

HCS Area level externalities These are basically blanket measures that will have an absolute effect upon all property prices within a HCS Area. The variables measuring the quality of local amenities were constructed using information from the CHCS, whilst ARC / INFO was used to calculate the proportion of each non-residential landuse. Housing density was calculated by similar means using the ADDRESS-POINT property coverage as a means of determining the number of properties in a particular HCS Area.

Community level externalities The social economic class variable computed using principal components analysis was used as a measure of social composition of each community, whilst prestige and desirability were captured using the community boundaries.

Conclusion

Table 4.8 is a summary of the locational variables constructed for the Inner Area. This illustrates the importance of the GIS in generating locational externality measures, and indicates a possible reason as to why some of the previous research has failed to capture the complexity of location upon property prices. The next chapter will begin to explore how location impacts upon property prices by attempting to capture the spatial dynamics of the Cardiff housing market. This will make use of the GIS constructed for the macro-scale study. Once a hedonic specification has been developed that can model the spatial variation in proper prices, then this will be used in chapter six to estimate the value of locational externalities in Table 4.8 in detail.

Table 4.8 Inner Area locational attributes

Property Level	Street Level (cont.)
Accessibility to CBD	Street quality 0-50m: Above Ave
Accessibility to M4 motorway	Street quality 0-50m: Good
Accessibility to railway stations	Street quality 50-100m: Poor
Proximity to hospitals	Street quality 50-100m: Below Ave
Proximity to sports centres	Street quality 50-100m: Above Ave
Proximity to community centres	Street quality 50-100m: Good
Proximity to institutional centres	Street quality 100-200m: Poor
Proximity to local shops	Street quality 100-200m: Below Ave
Proximity to primary schools	Street quality 100-200m: Above Ave
Proximity to secondary schools	Street quality 100-200m: Good
Proximity to Bute Park	Street non-residential landuse.
Proximity to parks / open space	School Catchment: Willows High
Proximity to light industrial landuse	School Catchment: Fitzalan High
Proximity to heavy industrial landuse	School Catchment: Cantonia High
Rail 0 -50m	School Catchment: Cathays High
Rail 50 - 100m	School Catchment: St Teilo's High
Rail 100 - 150m	**HCS Area Level**
Rail 150 - 200m	% Local Authority tenure
River 0 - 50m	% Open space
River 50 - 100m	% Non-residential landuse
River 100 - 150m	Housing density
River 150 - 200m	Quality of local shops
Street Level	Quality of local public transport
Road Type: Primary	Quality of local sport facilities
Road Type: Secondary	Quality of local parks
Road Type: Residential	Quality of community facilities
Road Type: Cul-de-sac / Close	**Neighbourhood Level**
Street quality 0-50m: Poor	Social economic class
Street quality 0-50m: Below Ave	

5 The Spatial Dynamics of an Urban Housing Market

Introduction

The aim of this chapter is to investigate the spatial dynamics of an urban housing market, using Cardiff as a case study. Implicit in this aim is an attempt to incorporate space into the hedonic house price model. By modelling housing market dynamics within a spatial framework, two key features of the housing market can be addressed. Firstly, that it rarely operate as a unified whole, but rather as a series of submarkets delimited by the housing stock (bundles of housing attributes) and location (geographical areas). Secondly, that the trade-off between housing and transport costs will result in a concave, negative rent gradient from the city centre outwards. Both of these concepts are fundamental to urban economic theory - the former is synonymous with housing market disequilibrium and the latter with the bid-rent process. Previous research has shown that empirical evidence concerning these processes is contradictory. The success to which the hedonic models explain the spatial dynamics of the housing market can be assessed by using diagnostic tests.

Variables and Statistical Tests

The variables used in the analysis are summarised in Table 5.1. To prevent rounding errors due to large numbers, and to facilitate interpretation of the models, the continuous independent variables were deviate around their means, so the models were estimated with regards to the typical property - an average sized (750 sq.-ft) mid-terrace house. In addition, the house price variable was transformed using the natural log. This is a common practice in hedonic studies (e.g. Jones & Bullen, 1993) and is used to deal with the technical problems of non-linearity and variance heterogeneity related to house size (Addair et al. 1996). The subsequent model estimates were transformed back into pounds and expressed as the difference over the

Table 5.1 Housing attributes used in the macro-study

Variable	Abbreviation
House Price	
Total Floor Area (sq.-ft.)	Floor Area
Dwelling Type	
End-Terraced	ET
Mid-Terraced	MT
Semi-Detached	SD
Detached	D
Flats in Converted Building	FCB
Purpose Built Flats	FPB
Maisonettes	M
Bungalow	B
End-Link	EL
Mid-Link	ML
Number of Bedrooms	Beds
Number of Recreation rooms	Recs
Number of Bathrooms	Baths
Number of Shower rooms	Showers
Full Central Heating	Full CH
Number of Garages	Garages
Off-Road Parking	ORP
Age: New	New
Age: Post 1964	Post 1964
Age: 1918 – 1964	1918-64
Age: Pre-1918	Pre-1918
Garden: None	Gdn: None
Garden: Less than 5 metres	Gdn: < 5m
Garden: 5 – 50 metres	Gdn: 5-50m
Garden: More than 50 metres	Gdn: > 50m
In need of modernisation	Needs Mods
Swimming Pool	Swm Pool
Conservatory	Con
Distance to CBD	Dist CBD
Social Class	Social
Housing Quality	H.Qual
Local Authority Tenure > 50%	LA > 50%

base price of the typical property.

A top-down regression building approach was used, which starts by including all the independent variables, and then discarding those that do not have a significant role in determining house price variation. The mechanics behind the process is to ensure that the assumptions underlying the hedonic model are not violated. The five key assumptions are:

1. that the relationship between the dependent and independent variables is linear
2. that there is no severe multicollinearity between the independent variables
3. that the fitting procedure has not be unduly influenced by unusual observations
4. that the errors are homoscedastic
5. that the errors are not autocorrelated.

If any of these assumptions are violated, the desired properties of the OLS estimates no longer hold, and action is needed to produce a satisfactory model. This is important since any violation of these properties indicates that the housing market has been incorrectly modelled.

The main guide for assessing each variable is to test whether its t-statistic is significant, which for large samples at a five percent level is a value greater than 1.96. The variance inflation factor (VIF) can be used to test for the presence of multicollinearity. A high VIF suggests collinearity, and as a general rule of thumb for standardised data, a VIF > 10 indicates harmful collinearity (Chatterjee and Price, 1977). Diagnostic tests can also be performed to avoid reliance on statistical summaries. Graphical diagnostics are an important, integral part of the model building process, and are among the most sophisticated diagnostic techniques available (Dunn, 1989). In particular, the models need to be checked for non-linearities, heteroscedasticity and unusual observations that may have an anomalous influence upon the regression parameters. Three principle diagnostic tests were used. Firstly, the residuals were examined to check for unusually large values; secondly, partial regression plots for each of the independent variables were analysed; and finally, joint regression diagnostic tests were performed upon the partial regression residuals.

Heteroscedasticity in the error term can be checked by plotting the residuals of the hedonic model against the predicted house prices. A non-random plot is indicative of heteroscedasticity, particularly if the residuals increase with the predicted values. However, a more accurate test for heteroscedasticity is the widely used Breusch-Pagan test. This tests the null hypotheses that the error variance is a linear combination of the variables. The general strength of the test is that it does not require prior knowledge of the functional form involved, and that there is a computationally convenient means of calculating the test statistics (Kennedy, 1985). The test statistic follows a chi-squared distribution with 1 degree of freedom, and if significant the presence of heteroscedasticity is assumed.

Spatial autocorrelation occurs if the errors are not independent across space. An elementary method of checking for spatial autocorrelation is by mapping the model's residuals and checking for spatial patterning. More formal statistical tests depend upon measuring the similarity of the residuals vis-à-vis their locations of which the most well known is the Moran Index (Cliff and Ord, 1981). The Moran Index uses a similarity measure based on the covariance of the residuals and also their spatial proximity. After analysing of the data, it was decided that spatial proximity was best described by a weight matrix based upon community membership. The Moran Index is positive when nearby errors tend to be similar, negative when they tend to be more dissimilar, and approximate zero when values are arranged randomly and independently in space.

To aid interpretation, the hedonic models are summarised in a standard format - see Table 5.2. The first column contains the variable coefficients which represent the attribute prices. In the next column are the standard errors of the coefficients. A t-statistic can be calculated by dividing each coefficient by its standard error, thus allowing the significance of each particular variable to be assessed. The third column contains the standardized coefficients. A standard coefficient of 0.5 implies that a one standard deviation change in the variable will result in a 0.5 standard deviation change in house price, thus allowing the relative importance of each attribute in the model to be assessed. The fourth column contains the variance inflation factors (VIFs) as a measurement of multicollinearity. A value of greater than 10 indicates that multicollinearity may be harmful. The final column contains the Breusch-Pagan test statistics calculated as a means of checking for heteroscedasticity. The standard error of the residuals and the coefficient of determination, adjusted for the number of variables in the model, are displayed below each model. For the sake of clarity, the variables that were not significant at a five percent level of have been removed from the following results tables.

Table 5.2 Model 5.1 The traditional hedonic specification

Predictor	Coefficient	Standard Error	Standard Coefficient	VIF	Breusch Pagan
Constant	44882	69.13	0.0557	*	*
Floor Area	35.10	1.11	0.544	2.0	15.9
SD	3184	782	0.061	2.0	0.004
D	16448	1257	0.219	2.3	0.792
B	15977	1607	0.125	1.3	0.055
Baths	6478	1186	0.089	1.4	1.90
Showers	4949	881	0.077	1.4	42.23
Full CH	4568	773	0.075	1.2	0.32
Garage	3146	566	0.081	1.7	0.07
ORP	2825	596	0.065	1.6	0.26
Gdn:None	-2926	758	-0.051	6.1	37.6
Gdn:5-50m	2931	755	0.067	5.7	13.1
Gdn:>50m	5519	1237	0.071	4.2	3.08
Needs Mods	-4628	11123	-0.053	1.1	0.307
Dist CBD	-1.80	0.170	-0.179	2.5	8.64
Social	4077.5	210	0.340	2.6	60.01
H.Qual	-1389	255	-0.0714	1.4	0.449
LA > 50%	3122	1323	0.035	1.9	65.43

s 13965			R-sq(adj)		83.4%

The Traditional Hedonic Specification

Introduction

This is the basic specification from which the subsequent specifications will be derived. It treats the housing market as unified, with supply and demand schedules held constant across urban space. Thus, there is no spatial variation in implicit prices. Also, the specification regards locational and structural attributes as operating independently, and at the same spatial level. Therefore:

$$P_i = \alpha\, X_i + \Sigma\, \beta_k\, S_{ki} + \Sigma \gamma_q\, L_{qi} + \varepsilon_i\, X_i \qquad\qquad 5.1$$

Where:
$i = 1, ..., N$ is the subscript denoting each property;
P_i is the price of property i;
$k = 1, ..., K$ is the number of structural attributes;
$q = 1, ..., Q$ is the number of locational attributes;
α, β, γ and ε are the corresponding parameters;
X_i is a column vector which consists entirely of ones.

The Basic Hedonic Model

Table 5.2 is a summary of the parameters of Model 5.1. The first thing to note is that all of the attributes have the theoretically correct signs, and the majority are significant at the one percent level. The R-squared statistic implies that the model is successful in explaining four-fifths of house price variation in Cardiff. Since the data have been deviated around their means, the constant term represents the average price of a typical property in Cardiff (a mid-terraced house with floor size of around 750 square feet), which is estimated as £44,882. The implicit prices of the attributes reveal some interesting results. As would be expected, the standardised coefficients suggest that the most influential attribute in determining house price is floor area, which has a marginal price of around £35.10 per square foot. Next is the socio-economic class variable, suggesting that locational attributes are fundamental in determining property prices.

The dwelling type attributes reflect the differences in price of different types of housing. The model suggests that there are no significant differences in price *ceteris paribus,* between the base category of mid-terrace housing, and linked properties, maisonettes, and flats, since these were insignificant at the five percent level and hence have been omitted. However, there are significant price differentials between terraced housing and the other dwelling types with detached houses and bungalows both having a notable influence increasing the price of a typical property by more than a third *ceteris paribus*. It is also interesting to note the relative importance of the detached dwelling variable. The standard coefficient suggests that its influence in the model is greater than any of the other attributes, with the exception of floor area and socio-economic class. Another unexpected result is the value of a separate shower room (£4,949), which is greater than central heating (£4,568) and a garage (£3,146). Finally, it is worth noting that the value of a property that is in need of modernisation is, on average, £4,628 less than one that is structurally sound. This compares to an estimated mean total repair cost of around £3,200 calculated in the CHCS.

The locational attributes appear to have more intuitive coefficients. Distance from the CBD is significant with the anticipated sign and is a fairly influential attribute of property price. Its implicit price suggests a rent gradient for Cardiff of £1.80 per metre, although this will be examined in more detail later. The standard coefficient of the socio-economic class variable would imply that locational attributes have a significant impact upon house price, relative to the other housing attributes. Since the variable is a principal component, the implicit price of £4,077 is harder to interpret. In areas of high socio-economic class (principal component score of between 1.58 to 3.7) locational effects add an extra £6,442 - £15,097 to the value of a typical property. Conversely, in areas of low socio-economic class (principal component score of between -1.28 to -6.74), locational effects reduce the value of a typical property by between £5,219 to £27,482. Housing quality operates upon house prices in a similar fashion, albeit to a lesser extent. The actual locational externalities that are responsible cannot be determined due to the surrogate nature of the variables, but shall be examined in more detail in the next chapter.

An unexpected result is the implicit price of the Local Authority housing tenure variable. The positive sign would suggest that properties located in areas of predominately Local Authority tenure are marginally more expensive (by £3,122) *ceteris paribus*. This is counter-intuitive since one would expect house prices to be lower due to the stigma element attached to these areas. Therefore, this result could suggest that house prices in these areas are inflated due to restricted availability of houses for sale. Since the variable indicates areas of predominately Local Authority tenure (over 50%), owner occupation would be low and hence the supply of properties constrained. If the demand for housing in such areas is large enough, this restriction in supply would result in an increase in house prices relative to the rest of the city. The alternative explanation is that the model has been misspecified.

The Breusch-Pagan test statistics indicate that several variables suffer from heteroscedasticity ($\chi^2 = 3.84$ at 95% with 1 degree of freedom). With respect to the structural attributes, floor area, the detached housing variable, the number of shower rooms and garden size all display heteroscedasticity. This may be indicative of spatial parameter drift, and omitted variables (Can, 1992), although preliminary data analysis concluded that floor area may also suffer from non-constant variance.

Variables that are not heteroscedastic are those attributes that are restricted to a small subset of housing bundles and locations, such as large gardens. The Moran test statistic (I=0.170) indicates that spatial autocorrelation is present (I = 0.074 at the 95% level), and hence the model over-predicts property prices in areas of smaller housing, lower socio-economic class and poor environmental quality and vice versa.

Housing Bundles and Structural Interactions

Introduction

The traditional specification asserts that the implicit price of a structural attribute is constant across bundles of housing attributes, such that the price for a unit of floor area is identical for terraced housing as for detached housing. However, Rosen (1974) has argued that these housing attribute bundles cannot be untied and repackaged to reflect the consumers desired mixed of attributes. This implies that the available mix of internal structural attributes of a housing bundle is limited and constrained. For instance, it is unusual for a terraced house to have more than four bedrooms, and a detached house to have less than three, although such a combination is theoretically allowed by the traditional specification. Since different types of households will desire different mixes of attributes, housing bundles may have different supply and demand schedules operating upon them. This may result in the variation of the implicit price of structural attributes between housing bundles. If this is the case, then using a single variable to measure the implicit price of an attribute may result in heteroscedasticity due to non-constant variance, since the single variable is measuring the effect of several (omitted) variables. In addition, the coefficient of the single variable will represent the weighted average of the implicit prices of the omitted attributes, and not the true implicit price of the attribute. However, with few exceptions (e.g. Schnare & Struyk, 1976), there has been very little work done on modelling housing bundles, despite the concept being apparently fundamental in the housing market literature.

To account for this, Model 5.1 was expanded to allow the distinct housing bundles within Cardiff to be explicitly modelled. Since it is usual to describe a property by its dwelling type, and each dwelling type embodies a typical set of structural attributes, this was used to categorise each housing bundle. In Model 5.1, the dwelling type variable is regarded as an additional premium on the price of a mid-terrace. But significantly higher implicit prices are commanded for detached housing and bungalows. This would be expected if the dwelling type dummy was capturing the effects of omitted structural variables. A similar argument can be applied to the estimated implicit prices for shower rooms, since this attribute tends only to be available within certain housing bundles. Hence, Model 5.1 was expanded by interacting the internal structural variables with the dwelling type dummy variables. These interaction terms represent the previous omitted variables. In addition, the continuous variables measuring the number of bathrooms and garages were converted into several dummy variables, since there is no reason to expect the implicit price of these attributes to increase at a constant rate. Since no explicit spatial variables were included in the expansion equations, the resulting model was called the traditional structural expansion model.

Model 5.2 - The Structurally Expanded Specification

Table 5.3 is a summary of the parameters of Model 5.2. As hypothesized, the implicit prices of the internal structural attributes varied with dwelling type. Generally, the structural attributes for terraced properties, linked properties, maisonettes and flats all had similar implicit prices, and did not significantly vary with respect to each other, and have been omitted. This suggests that the structural attribute mix for these housing bundles are very similar. Thus the implicit price of floor area in all these properties is £36.60 per square foot. The model illustrates how this value varies between the remaining dwelling types. For instance, the model implies that the price of floor area in detached housing and bungalows are a third and a half more expensive respectively ((£36.60 + £13.0) = £49.60 and (£36.60 + £18.70) = £55.30 per square foot.). The variance inflation factors (VIFs) for each variable illustrate that the addition of the floor area interaction terms has not lead to a noticeable increase in multicollinearity in the model.

The standard coefficients suggest that floor area is still the most influential attribute in explaining the variation in house prices, although this has now been distributed between floor area in detached houses and bungalows. A surprising result is the negative implicit price for bungalows with two bathrooms. This is counter-intuitive since it suggests that bungalows with two bathrooms are significantly cheaper (by approximately £10,900) than bungalows with one bathroom. Unless there is a structural explanation (such as two bathroom bungalows being located in less desirable areas, or are smaller in overall size), these results imply that the model is misspecified.

Table 5.3 Model 5.2 The structurally expanded specification

Predictor	Coefficient	Standard Error	Standard Coefficient	VIF	Breusch Pagan
Constant	47425	79.36	0.051	*	*
Floor Area	36.60	1.25	0.535	2.1	14.2
Floor SD	2.90	0.99	0.053	2.2	1.62
Floor D	13.0	1.45	0.203	3.4	0.76
Floor B	18.70	1.95	0.139	1.6	1.02
Full CH	5689	767	0.083	1.2	0.05
Garage	4025	606	0.095	1.7	0.005
ORP	2947	645	0.064	1.6	0.043
D Shower 1	6827	1712	0.060	1.7	2.08
FPB Shower 1	15978	3234	0.05	1.1	13.4
D Baths 2	7337	2556	0.041	1.4	0.49
B Baths 2	-10921	4044	0.034	1.2	0.91
Gdn:None	-4238	811.9	0.071	1.5	34.4
Gdn:5-50m	3343	804	0.075	2.5	18.71
Gdn:>50m	5280	1354	0.065	2.0	7.76
Needs Mod	-5379	1537	0.055	1.1	0.19
Dist CBD	-1.90	0.19	0.172	2.4	5.4
Social	4298	225	0.337	2.6	48.9
H.Qual	-1504	274	0.072	1.4	0.8
LA > 50%	2789	1416	0.03	1.9	51.0

s   13356   R-sq (adj)   85.4%

The locational attributes are still highly influential in the model. The social class of an area is still the second most important attribute influencing house price, whilst the accessibility attribute now suggests a rent gradient of £1.90 per metre. However, the Local Authority tenure variable still predicts an additional premium for housing in areas of predominately Local Authority owned stock. Finally, the overall standard error has been reduced, indicating that the model is a better predictor of house price than in Model 5.1.

The Breusch-Pagan test statistics reveal that heteroscedasticity is still prevalent in the floor area attribute, and has not decreased significantly with respect to the previous model. However, the shower room attribute has seen a marked reduction in heteroscedasticity, and is now only problematic with respect to purpose built flats. The introduction of the structural attribute interaction terms has increased heteroscedasticity in the garden size attributes, although it has had a marked reduction in the locational attributes. Spatial autocorrelation is still a problem, although the Moran Index indicates a decrease ($I = 0.167$) compared to Model 5.1. These results suggest that it was correct to re-specify the model, and that differentiation between housing bundles is an important structural feature of the Cardiff housing market. However, the existence of residual heteroscedasticity and spatial autocorrelation suggests that spatial submarkets may be present causing spatial parameter drift to occur.

Spatially Expanding the Fixed Parameters

The Spatial Parameter Drift Specification

The spatial parameter drift specification models the spatial submarkets by expanding the fixed parameters of the traditional hedonic specification. Specifically:

$$P_i = \Sigma\,(\alpha_0 + \alpha_1 Z)\,X_i + \Sigma\,(\beta_{k0} + \beta_{k1} Z)\,S_{ki} +$$
$$\Sigma\,(\gamma_{q0} + \gamma_{q1} Z)\,L_{qi} + \varepsilon_i\,X_i \qquad\qquad 5.2$$

Where Z is a measure of location.

This specification is similar to the discrete space expansion equation (equation. 2.19) and can be regarded as interacting structural attributes with a measure of location. It is hypothesized that it was the omission of these interaction terms that was the cause of the residual heteroscedasticity in the traditional specification. The specification is operationalised by permitting the implicit prices of the housing attributes to interact with the social class variable (see equation. 2.21). In such a specification, there is no implicit price for social class *per se*. Instead, social class can be conceived as driving the implicit prices of the structural attributes across space. Social class was chosen since it can be argued that the valuation of a bundle of housing attributes will be determined in part by the income of the buyer, and this can be proxied by the social class variable. Moreover, since social class is also a proxy for locational attributes, this specification will illustrate how structural attributes vary with locational context.

It may be hypothesized that implicit prices of structural attributes will be more expensive in areas of relatively high socio-economic class *ceteris paribus,* than in areas of lower socio-economic class. This is supported by the significance of the social class variable in determining house price in the earlier models. Two expansion equations are specified: a linear equation implying that changes in implicit prices are proportional to changes in socio-economic class.

$$\delta_k = \delta_{k0} + \Sigma \, \delta_{k1j} \, Z \qquad\qquad 5.3$$

Where Z is now social class.
δ represents the fixed parameters in the model.

a non-linear quadratic equation implying that the greatest shifts in attribute implicit prices occurs in areas of very high or very low socio-economic class.

$$\delta_k = \delta_{k0} + \Sigma \, \delta_{k1j}Z + \Sigma \, \delta_{k2j}Z^2 \qquad\qquad 5.4$$

For reasons of clarity, insignificant drift variables have been omitted from the subsequent results tables.

The Traditional Spatial Drift Specification

Model 5.3 The linear model The fixed parameters in Model 5.1 were expanded using the linear expansion equation (equation 5.3), and the model was re-estimated. The average price of the typical property was estimated as £46,812, and this now varies by the social class of an area (Table 5.4). In areas of above average social class (1.58 - 3.7), this price increases by around £6,647, whilst in areas of below average social class (-1.28 - -3.7), it is around £10,000 cheaper. This can be used to estimate the additional price of locating in a specific community. Table 5.5 is a summary of the social class measures for each community. Using these values, the average price of a typical property can be allowed to vary. This

Table 5.4 Model 5.3 The linear spatial parameter drift specification

Predictor	Coefficient	Standard Error	Standard Coefficient	VIF	Breusch Pagan
Constant	46812	73.84	0.06	*	*
Z.Constant	2518	468.9	0.20	1.0	9.1
Floor Area	35.58	1.12	0.529	1.9	21.32
Z.Floor Area	2.99	0.61	0.206	3.2	2.63
SD	2637	782	0.04	2.0	0.017
Z.SD	740	338	0.04	2.5	1.3
D	18393	1339.6	0.23	1.7	0.49
B	16371	1627	0.12	1.4	0.79
Baths	8365	1262	0.11	1.5	4.3
Z.Baths	-2814	575.5	-0.09	1.4	0.013
Showers	5858	934	0.09	1.6	11.86
Z.Showers	-2225	407.5	-0.10	1.9	0.164
Full CH	4501	721	0.07	1.2	0.01
Garage	3234	581	0.08	1.8	1.23
Z.Garage	-618	273	-0.04	2.4	9.1
ORP	2228	599	0.06	1.6	6.14
Gdn:None	-4144	770	-0.08	1.6	12.6
Gdn:5-50m	2746	754	0.06	2.5	6.97
Gdn:>50m	5989	1248	0.07	2.0	2.15
Needs Mods	-4366	1119	-0.05	1.2	0.005
Dist CBD	-1.67	0.176	-0.13	2.5	1.89
H.Qual	-1524	266	-0.05	1.5	0.51
La > 50%	-10220	2542	-0.14	9.5	2.69
Z.La > 50%	-4370	812	-0.17	8.5	3.3

s   9474   R-sq(adj)   84.5%

suggests that the most expensive communities are Radyr & St. Fagans, Lisvane & St. Mellons and Cyncoed, with the least expensive being Butetown, Adamsdown and Ely. The relatively large standard coefficient suggests that the way the average price of a property varies with social class is an important feature of the model. Floor area remains the most important attribute in the model (standard coefficient of 0.529).In areas of average social class, the price of floor area is estimated as £35.58 per square foot. This then varies by £2.99 per square foot as social class

Table 5.5 Communities ranked by social class

Community	Social Class	Community	Social Class
Butetown	-5.202	Gabalfa	0.162
Adamsdown	-2.833	Canton	0.415
Ely	-2.643	Rumney	0.455
Riverside	-2.441	Whitchurch &	1.468
Grangetown	-1.972	Tongwynlais	
Plasnewydd	-1.909	Roath	1.517
Splott	-1.693	Llanishen	1.910
Trowbridge	-1.365	Heath	2.320
Llanrumney	-1.153	Llandaff	2.494
Caerau	-1.144	Rhiwbina	2.867
Cathays	-0.756	Cyncoed	2.996
Llandaff North	-0.450	Lisvane & St Mellons	3.305
Fairwater	-0.021	Radyr &	3.528
Pentwyn	0.145	St Fagans	

deviates from this average. Hence, in areas of relatively high social class, a household would have to pay on average an additional £8.72 per square foot. Conversely, in areas of relatively low social class, the price would be £7.73 per square foot cheaper. Detached housing and bungalows still command a high premium, although this premium does not vary, but remains stable across urban space. However, the premium for semi-detached houses is dependent upon social class, with the price of a semi-detached house increasing as the social class of an area increases. This instability can be explained by the ubiquitous nature of semi-detached houses compared to detached houses and bungalows. The latter tend to be concentrated in areas of similar social class, and hence their implicit prices are less likely to drift.

The only other structural attribute prices that experience drift are baths, showers and garages. In all three cases, the drift coefficient are negative, implying that the additional value of such attributes are increasingly higher in areas of lower social class. This rather counter-intuitive result may be explained by the fact that attributes such as two bathrooms and double garages are rarer in the housing stock in these areas, and hence with sufficient demand, their restricted supply may increase their value. However, it may be the case that additional structural attributes are valued less in areas of higher social class in favour of locational attributes.

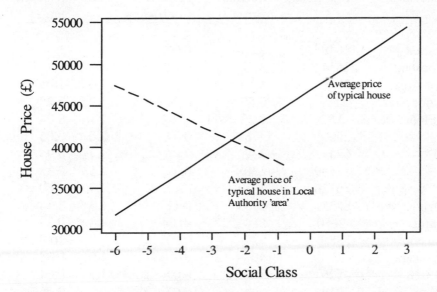

Figure 5.1 Local Authority housing stock and social class: model 5.3

An interesting result is the implicit price of properties in areas of predominately Local Authority owned housing stock. The average implicit price is negative, as is expected. This price then varies depending upon the social class of the area. Therefore, in areas of average social class, Local Authority housing stock has a negative effect of around £10,000. This negative effect varies with social class. This is summarised in Figure 5.1, and shows that, for low social class (-6 to -3), properties in areas of predominately Local Authority tenure are more desirable and hence more expensive, than similar properties in non-Local Authority areas. This may be explained by the fact that a high proportion of Local Authority dwellings in Cardiff were built in the inter-war period when construction standards were high (Short, 1981). In particular, they tend to be of better structural quality and built to a lower density than comparable properties in other areas of low social class, which are synonymous with streets of inner city terraces. This combination of structural quality and locational externalities may have resulted in the 'stigma effect' only becoming significant in areas of slightly below average (greater than -3) social class.

**Table 5.6 Model 5.4 The non-linear spatial parameter drift
          specification**

Predictor	Coefficient	Standard Error	Standard Coefficient	VIF	Breusch Pagan
Constant	46486	72.46	0.06	*	*
Z.Constant	2614	470	0.21	1.0	9.7
Floor Area	36.77	1.29	0.54	2.6	18.53
Z.Floor Area	2.59	0.59	0.202	5.7	3.28
ZZ.Floor Area	-0.52	0.23	-0.04	2.1	0.04
SD	2560	775.6	0.04	2.0	0.05
Z.SD	674	337	0.04	1.7	0.94
D	17839	1308	0.22	2.5	0.33
B	16126	1614	0.12	1.4	0.69
Baths	8235	1253	0.11	1.5	3.34
Z.Baths	-2650	584	-0.04	1.5	0.34
Showers	5885	927	0.09	1.6	10.9
Z.Showers	-2170	407	-0.10	1.9	0.38
Full CH	4526	717	0.07	1.2	0.01
Garage	2919	564	0.07	1.7	0.66
ORP	2335	594	0.06	1.6	7.1
Gdn:None	-4060	765	-0.07	1.6	13.14
Gdn:5-50m	2737	750	0.06	2.5	5.63
Gdn:>50m	6032	1241	0.07	2.0	1.75
Needs Mods	-4398	1111	-0.05	1.2	0.001
Dist CBD	-1.69	0.175	-0.13	2.5	1.64
H.Qual	-1446	262	-0.08	1.5	0.16
La > 50%	-9281	2543	-0.13	9.4	3.2
Z.La > 50%	-3826	802	-0.15	8.1	3.7

s	9408		R-sq(adj)		84.50%

In these areas, Local Authority housing stock may be viewed with more disdain by the more relatively more affluent purchasers, who will also have a better choice of housing stock of comparable quality and size.

Model 5.4 The non-linear model This model presupposes that housing attributes interact with social class in a complex, non-linear fashion. Model 5.3 was re-estimated using the quadratic expansion equation and is summarised in Table 5.6. The principal result of the model is the implication that the spatial variation of floor area is non-linear. The model suggests that not only does floor area vary with social class, but this variation is greater in areas of either very high and very low social class. In both cases, this results in a deduction of £0.52 per square foot. In addition, the garage variables now become stable across the housing market, suggesting that the spatial variation in the linear model was caused by the number of garages acting as a proxy for unaccounted spatial variation in floor area. Finally there is very little difference between the standard errors in the two models, suggesting that the non-linear model does not significantly explain more of the variation in house prices. The remaining variables continue to be robust.

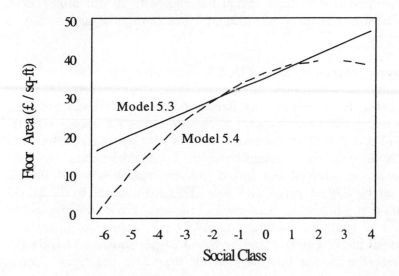

Figure 5.2 Price per unit floor area by social class: model 5.3 and model 5.4

Figure 5.2 illustrates the way in which the unit price of floor area varies with social class in Models 5.3 and 5.4. It is the non-linear plot which is of interest. This suggests that the unit price of floor area increases at a decreasing rate until it peaks at around £40 per square foot in areas of slightly above average social class (between 1 and 2). The price per square foot then starts to decrease. This result is interesting since it suggests that higher income purchasers may place marginally less value upon the structural attributes of housing than lower income purchasers. The corollary of this is that higher income purchasers may be placing marginally greater value upon locational attributes. This is supported by the way the bathrooms and shower room parameters drift, being progressively less expensive in areas of higher socio-economic status.

Testing for spatial effects Both models suffer from heteroscedasticity, although this has decreased significantly compared to model 5.1, the non-expanded equivalent. The linear model is more heteroscedastic, with heteroscedasticity concentrated in the floor areas variables, and related attributes such as baths, showers and garden size. This is caused in part by the omitted structural interaction variables, although the garden size variables suggest that perhaps spatial heterogeneity is still problematic. The significant Moran test ($I = 0.165$) indicates that spatial autocorrelation is still present.

The Structurally Expanded Spatial Drift Specification

Model 5.5 The linear model The fixed parameters in Model 5.2 were expanded using the linear expansion equation and the model was re-estimated (Table 5.7). The new model reveals how the internal structural attributes of the different dwellings types drifts with social class. The price of floor area for terraced and linked houses, maisonettes and flats is estimated to be £39.33 per square foot. This then varies by £2.91 per square foot, depending upon social class of the area. Floor area in detached housing has an additional premium of £14.24 per square foot in areas of average social class, with this varying by £3.24 per square foot, although the negative sign implies that the price of floor area per square foot is progressively cheaper as social class increases. Floor area in bungalows also command an additional premium of £14.42 per square foot, although this premium does not drift but rather is constant across Cardiff. This can be explained by the fact that bungalows are concentrated within specific communities which have very similar social class. Figure 5.3 summarises this interaction of floor area and social class for each property type. It is interesting to note the negative slope for detached houses and the implication that in areas of high social class, there is very little difference in the price per unit floor area between housing types, with the exception of bungalows. These are the most expensive dwelling type per square foot in areas of above average social class.

Table 5.7 Model 5.5 The structurally expanded linear spatial parameter drift specification

Predictor	Coefficient	Standard Error	Standard Coefficient	VIF	Breusch Pagan
Constant	49518	81.40	0.04	*	*
Z.Constant	2664	548	0.204	1.1	10.78
Floor Area	39.33	1.16	0.55	1.8	15.4
Z.Floor Area	2.91	0.67	0.19	7.3	2.79
Floor D	14.34	1.61	0.22	4.1	0.35
Z.Floor D	-3.24	0.69	-0.11	3.5	0.00
Floor B	14.42	1.83	0.101	1.4	0.155
D Bath2	26210	7383	0.12	8.3	0.08
Z.D Bath2	-5173	2210	0.08	8.6	0.055
Z.B Bath2	-5652	2070	-0.04	1.2	0.29
D Shower1	7458	1703	0.06	1.7	1.94
Z.FPB Shower1	-4451	849	-0.06	1.4	0.38
Z.SD Shower1	-3569	1399	0.04	1.1	2.15
Full CH	5332	782	0.08	1.2	0.02
Garage	4167	605	0.10	1.7	0.02
ORP	2979	637	0.06	1.5	1.5
Gdn:None	-4977	825	-0.08	1.6	17.3
Gdn:5-50m	4011	755	0.07	2.1	9.58
Gdn:>50m	6737	1296	0.08	1.8	4.78
Needs Mods	-4932	1207	-0.05	1.1	0.12
Dist CBD	-1.83	0.19	-0.16	2.5	1.9
H.Qual	-1849	284	-0.08	1.5	0.82
La > 50%	-11510	2774	-0.13	9.8	5.27
Z.La > 50%	-4887	900	-0.18	9.1	10.01

s	10230		R-sq(adj)		83.9%

The remaining structural attributes reveal an interesting geography of spatial drift. The implicit price of two bathrooms in a detached property is more expensive in areas of lower social class, whilst the implicit price of a two bathroom bungalow drifts such that they are cheaper in more affluent areas. In the previous, unexpanded model (Model 5.2), this variable had the wrong sign. It would now seem that this was caused by misspecification due to omitted variable bias. In a similar fashion, shower rooms in purpose

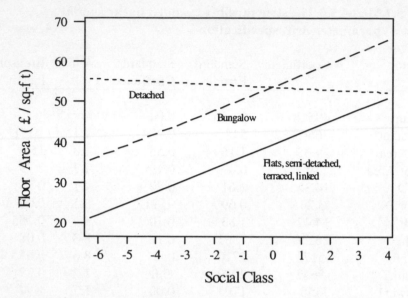

Figure 5.3 Price per unit floor area by social class: model 5.5

built flats and semi-detached housing are also cheaper in more affluent areas, whilst those in detached houses are stable across the housing market. The remaining structural attributes are invariant, whilst Local Authority dwelling behave in a similar fashion to the previous model.

Model 5.6 The non-linear model Similar to Model 5.4, floor area was the only variable to drift across space in a non-linear fashion (Table 5.8). This additional parameter has the effect of increasing the spatially varying implicit price of floor area in bungalows, but reducing the implicit price of floor area in all other properties. The remaining parameters appear to be robust with respect to non-linear spatial drift. Figure 5.4 summarises the new patterns of floor area drift. The model now suggests that the implicit of price floor area in all dwelling types increases with social class at a decreasing rate, before peaking and then declining in areas of above average social class. The steepest decline is for detached housing, in which floor area is the most expensive in areas of average social class. Again, bungalows are the most expensive properties with respect to size in areas of above average social class.

Figure 5.5 shows the same graph re-plotted using average community level social class values. This suggests that, for detached housing, the most expensive housing stock is located in Cathays, Llandaff

North and Fairwater, whilst buyers are paying less for housing space in Cyncoed, Lisvane & St. Mellons and Radyr & St Fagans. This may imply that buyers in these areas value locational attributes more than in areas of average social class.

Table 5.8 Model 5.6 The structurally expanded non-linear spatial parameter drift specification

Predictor	Coefficient	Standard Error	Standard Coefficient	VIF	Breusch Pagan
Constant	49222	81.10	0.05	*	*
Z.Constant	3106	560	0.24	1.1	12.2
Floor Area	41.34	1.35	0.58	2.5	12.1
Z.Floor Area	2.42	0.68	0.16	8.5	5.3
ZZ.Floor Area	-0.55	0.165	-0.06	2.4	0.69
Floor D	13.9	1.59	0.21	4.2	0.25
Z.Floor D	-2.50	0.71	-0.08	3.8	1.8
Floor B	14.9	1.82	0.11	1.4	0.21
D Bath2	22067	7141	0.11	8.5	0.27
Z.D Bath2	-3807	1667	0.06	8.9	0.4
Z.B Bath2	-5409	2048	-0.03	1.2	0.17
D Shower1	7487	1682	0.06	1.7	1.32
Z.FPB Shower1	-4286	840	-0.06	1.4	0.15
Z.SD Shower1	-3309	1390	-0.04	1.1	2.17
Full CH	5318	773	0.08	1.2	0.08
Garage	4041	597	0.09	1.7	0.065
ORP	2934	628.3	0.06	1.5	1.8
Gdn:None	-4809	816	-0.07	1.6	18.7
Gdn:5-50m	3945	746	0.08	2.1	10.21
Gdn:>50m	6929	1283	0.08	1.8	5.7
Needs Mods	-4902	1187	-0.05	1.1	0.07
Dist CBD	-1.87	0.19	-0.164	2.5	1.65
H.Qual	-1764	281	-0.08	1.5	0.28
La > 50%	-10448	2786	-0.12	9.8	6.9
Z.La > 50%	-4326	907	-0.17	9.2	11.8

s	10127		R-sq(adj)		84.0%

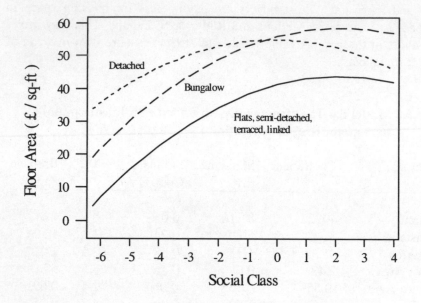

Figure 5.4 Price per unit floor area by social class: model 5.6

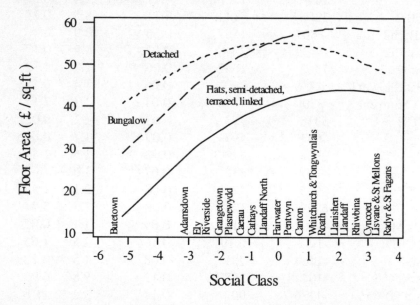

**Figure 5.5 Price per unit floor area by community level social class:
model 5.6**

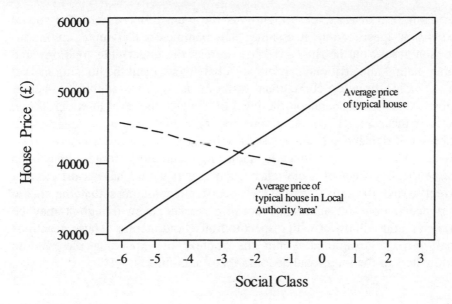

Figure 5.6 Local Authority housing stock and social class: model 5.6

Finally, Figure 5.6 is a summary of the affect that an area with a high degree of Local Authority tenure has upon property prices. The graph is very similar to Figure 5.1, implying that the results are robust to changes in model specification. Again it would appear that, in areas of low social class, properties are more desirable if they are located in areas of predominately Local Authority tenure, with this desirability declining with increasing social class.

Testing for spatial effects Heteroscedasticity was reduced in the floor area variable in both models compared to Models 5.3 and 5.4. However, there was an increase in heteroscedasticity in the garden and Local Authority variables. This heteroscedasticity was slightly greater in the non-linear model. However, these did display a marked reduction in heteroscedasticity compared to Model 5.2, suggesting that spatial heterogeneity was a problem in the traditional specifications. Finally, positive spatial autocorrelation is still prevalent in Model 5.6, with a significant Moran Test ($I = 0.157$), despite the attempt to capture submarket processes. Furthermore, residual maps reveal a strong community clustering of values, which remain substantially unchanged, despite the attempt to explain spatial variations. This distinct clustering of residuals into discrete areas highlights the main problem with the spatial expansion specification. It assumes that parameters drift in a continuous fashion across the housing market, whereas the underlying topology and other natural and artificial barriers are likely to segment the housing market in a more discrete, contiguous fashion. It is also doubtful whether submarkets would operate at the level of the ED, the level at which social class is measured. Urban economic theory would appear to suggest that supply and demand schedules would only significantly differ across wider areas, such as communities. In addition, community context may also influence the affect of social class such that it would have a differential effect across the city. Finally, the specification assumes that the model variance is constant across the housing market, even though it may be argued that it will differ with respect to household income. These questions shall now be examined within the context of expanding the random parameters of the traditional hedonic specification.

Spatially Expanding the Random Parameters

The Multi-level Specification

Introduction The previous models were estimated at the level of the individual property, although the locational attributes were aggregated at the level of the ED. The multi-level specification requires the different spatial levels to be explicitly modelled: the individual property, the ED and the community. The ED level captures the local variations that approximate individual streets, and are contextualised by the locational attributes. The community level approximates submarkets, at which housing attributes are deemed to vary. These three spatial levels were incorporated into the traditional hedonic specification by expanding the random parameters. Thus:

$$P_{ijk} = \alpha_{jk} X_{ijk} + \Sigma \beta_m Z_{mijk} + \varepsilon_{ijk} X_{ijk} \qquad 5.5$$

where:
Z is the vector of m housing attributes,
$k = 1, ..., 26$ is the number of communities,
$j = 1, ..., 453$ is the number of enumeration districts
$i = 1, 1500$ is the number of properties.

This is expanded using the equation:

$$\alpha_{jk} = \alpha + \mu\alpha_{jk} + \mu\alpha_k \qquad 5.6$$

to form the random intercepts model:

$$P_{ijk} = \alpha_{jk} X_{ijk} + \Sigma \beta_m Z_{mijk} + (\mu\alpha_{jk}X_{ijk} + \mu\alpha_k X_{ijk} + \varepsilon_{ijk} X_{ijk}) \qquad 5.7$$

    This specification has several conceptual and technical advantages over the previous specifications. The multi-level concept distinguishes between the compositional effects of the housing stock (i.e. the structural attributes) and the contextual effects of the street and neighbourhood (i.e. the locational attributes). The previous specifications have treated structural and locational attributes as having an equal effect upon house price, even though the latter are typically shared amongst many properties. This leads to the problem of confounding the affects of structural attributes with locational attributes. By taking into account the context of location, the multi-level specification will allow not only the structural attributes to vary across space, but also the locational attributes, ameliorating the problem of the spatial drift specification in which the social class effect is constant across the city, regardless of neighbourhood context. Also, unlike the previous specifications, the multi-level specification also has the added technical advantage of being able to handle inherently spatial data, resulting in the implicit modelling of spatial autocorrelation and spatial heteroscedasticity. The modelling was undertaken using the multi-level modelling package Mln (Rasbash et al, 1995).

Model 5.7 The grand mean model This is the simplest model, since it contains no fixed terms except for the overall intercept. This estimates the average house price for the whole of Cardiff, and gives a figure of £57,000 (Table 5.9), which compares to £56,000 for the ecological mean. The model allows the variation around this grand mean to be decomposed into variation at the level of the individual property, ED and community. The greatest variation occurs between individual houses within a community, although over a third of the variation occurs between communities. This is interesting since it suggests that the attribute prices vary significantly between places, which could be indicative of submarkets.

    The significance of the parameters model may be judged by two methods (Woodhouse et al, 1996). The first is to calculate a t-statistic. This

Table 5.9 Model 5.7 The grand mean model

FIXED

Predictor	Coefficient	Standard Error
Constant	56954	3449

RANDOM

Parameter	Coefficient	Standard Error
Community Level		
Constant	4991	1485
ED Level		
Constant	2562	338
Property Level		
Constant	5739	257

$-2*(\text{log-likelihood}) = 1142.35$

works well for fixed parameters, and is similar to significance testing in OLS regression. However, for random parameters, the distribution of the t-statistic may depart considerably from normality, especially in small samples. Instead, Woodhouse recommends using a likelihood ratio statistic. By checking if the likelihood ratios of successive models are significantly different, the significance of the additional terms in the model may be evaluated.

Table 5.10 describes this variation in price around the grand mean at the community level. It is estimated that houses in Lisvane and St Mellons are some £40,000 more expensive than the average Cardiff price, whilst houses in Adamsdown are nearly £20,000 cheaper. This is illustrated graphically in Figure 5.7. Here, the cheapest communities are concentrated within the Inner Area, and peripheral estates with the highest premiums found in the northern suburbs. As Jones & Bullen (1994) have commented, the model has merely estimated an overall average for Cardiff, and individual averages for each community, and has therefore reproduced ecological estimates. However, these are precision-weighted, so that the estimate for the community effect based on a small number of sales are

Table 5.10 Community level premiums

Model 5.7	Price	Model 5.8	Price	Model 5.9	Price
Lisvane & St Mellons	40700	Roath	16050	Roath	8759
Radyr & St Fagans	31158	Cyncoed	12958	Riverside	8223
Roath	30295	Llandaff	9058	Cyncoed	7228
Cyncoed	27296	Butetown	8490	Butetown	6620
Heath	20442	Heath	7985	Cathays	6364
Llandaff	19213	Cathays	7765	Plasnewydd	5782
Rhiwbina	18195	Riverside	7601	Llandaff	3857
Whitchurch & Tongwynlais	15114	Plasnewydd	6347	Heath	3263
Llanishen	14174	Canton	5268	Canton	2826
Canton	1003	Whitchurch & Tongwynlais	4580	Whitchurch & Tongwynlais	2722
Llandaff North	-1628	Rhiwbina	4224	Llanishen	1827
Rumney	-1714	Gabalfa	2114	Rhiwbina	1825
Gabalfa	-2700	Llanishen	1523	Llandaff North	1384
Plasnewydd	-4241	Radyr & St Fagans	152.7	Gabalfa	763.8
Butetown	-4267	Lisvane & St Mellons	-335.6	Lisvane & St Mellons	-233
Riverside	-5406	Llandaff North	-375	Adamsdown	-639
Cathays	-7627	Splott	-2381	Radyr & St Fagans	-1153
Trowbridge	-9977	Adamsdown	-2408	Splott	-1198
Fairwater	-10349	Grangetown	-2762	Grangetown	-1923
Grangetown	-12060	Rumney	-4043	Fairwater	-3250
Ely	-14567	Fairwater	-4779	Rumney	-3397
Pentwyn	-14871	Pentwyn	-9904	Trowbridge	-7141
Caerau	-15497	Trowbridge	-11313	Llanrumney	-7614
Llanrumney	-15812	Caerau	-11850	Ely	-8504
Splott	-16982	Llanrumney	-12285	Caerau	-8829
Adamsdown	-19176	Ely	-13143	Pentwyn	-9212

Table 5.10 (cont) Community level premiums

Model 5.10	Price	Model 5.11	Price	Model 5.12	Price
Roath	8314	Riverside	10481	Riverside	9982
Riverside	7723	Roath	8706	Roath	7913
Cyncoed	7019	Cyncoed	5471	Cyncoed	5749
Butetown	6131	Llandaff	4865	Plasnewydd	4575
Cathays	5899	Plasnewydd	4681	Llandaff	4468
Plasnewydd	5420	Cathays	3836	Cathays	4025
Llandaff	4332	Heath	3350	Whitchurch & Tongwynlais	3331
Heath	3657	Canton	3270	Heath	3252
Whitchurch & Tongwynlais	3058	Whitchurch & Tongwynlais	3037	Llanishen	2835
Canton	2855	Rhiwbina	2591	Rhiwbina	2684
Llanishen	2486	Butetown	2373	Canton	2611
Rhiwbina	2334	Llanishen	2248	Llandaff North	1946
Llandaff North	1363	Llandaff North	1758	Butetown	1864
Lisvane & St Mellons	809	Lisvane & St Mellons	811	Lisvane & St Mellons	1325
Gabalfa	531	Gabalfa	445	Radyr & St Fagans	511
Radyr & St Fagans	-376	Radyr & St Fagans	437	Gabalfa	233
Grangetown	-2160	Grangetown	-1695	Grangetown	-2098
Adamsdown	-2206	Fairwater	-2646	Fairwater	-2599
Splott	-2662	Rumney	-2772	Rumney	-2761
Fairwater	-3041	Splott	-3673	Splott	-3804
Rumney	-3250	Adamsdown	-4552	Adamsdown	-4030
Trowbridge	-6864	Llanrumney	-5999	Llanrumney	-5857
Llanrumney	-7257	Trowbridge	-6210	Trowbridge	-5953
Ely	-8302	Ely	-6406	Ely	-6191
Caerau	-8624	Caerau	-8407	Caerau	-8446
Pentwyn	-9233	Pentwyn	-8818	Pentwyn	-8789

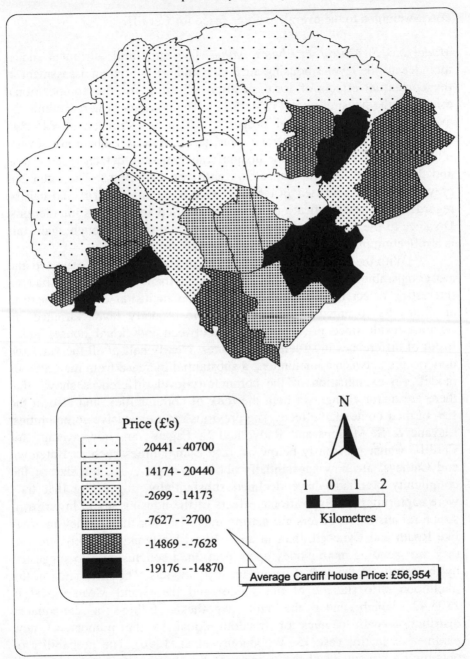

Price (£'s)

⠄⠄⠄⠄	20441 - 40700
⠒⠒⠒⠒	14174 - 20440
⣿⣿⣿	-2699 - 14173
▦▦▦	-7627 - -2700
▨▨▨	-14869 - -7628
■■■	-19176 - -14870

N

1 0 1 2
Kilometres

Average Cardiff House Price: £56,954

Figure 5.7 Differential community level prices: model 5.7

down-weighted to the overall average price for Cardiff.

Model 5.8 Structural attributes model The structural attributes model includes all the level-1 housing attributes and will allow an assessment of the contextual effects of location after adjusting for the compositional effects of the housing stock (Table 5.11). The constant now represents the average price of the typical property, which is estimated at £45,000, reflecting the price of an average sized terraced house. There are several interesting features, notably the insignificance of semi-detached floor area, and the negative sign for bungalows with two bathrooms. The latter has previously been discussed in terms of spatial parameter drift and can be regarded as a misspecification caused by omitted social class interactions. Distance to the CBD also has a counter-intuitive sign, although again this is a reflection of omitted higher level interactions.

With these exceptions, the fixed term estimates are as expected and are comparable to the previous models, but their inclusion has had an interesting effect on the random terms. The inclusion of the structural attributes has resulted in a decline of the property level variance, an obvious result since price differences between individual houses are a result of differences in structural attributes. Nearly half of all the variation now occurs between communities, a substantial increase from the previous model. An examination of the community-level differences shows that there are major changes in both the rank of communities, and also in the size of their contextual effects. The previous most expensive communities, Lisvane & St Mellons and Radyr and St Fagans, are now average for Cardiff, whilst previously below average communities such as Butetown and Cathays, are now substantially above average. Also, the size of the community premiums have declined substantially, suggesting that they were capturing the compositional effects of the housing stock. In terms of structural attributes, buyers are getting much less for their money in areas like Roath and Cyncoed, than in areas like Llanrumney and Ely. Instead they are spending more money upon location, a conclusion also suggested by the results of the spatial parameter drift models. The difference in the likelihood ratio statistic of this model and the Grand Mean model is 1803.42, which under the null hypothesis follows a chi-squared distribution with degrees of freedom equal to the number of new parameters, in this case 15 (Woodhouse et al, 1996). The probability of obtaining a chi-square of this magnitude by chance is exceedingly small (less than 0.001), strongly indicating that the structural attributes have an important effect in explaining house price variation in the model.

Table 5.11 Model 5.8 The structural attributes model

FIXED

Predictor	Coefficient	Standard Error
Constant	44801	2321
Floor Area	36.11	1.69
Floor SD	-5.59	5.29
Floor D	3.56	0.28
Floor B	2.66	0.39
D Bath 2	5340	768
B Bath 2	-1532	615
D Shower 1	2076	445
Full CH	4092	924
Garage	5651	658
ORP	4244	804
Gdn: None	-4650	2114
Gdn: 5-50m	4443	859
Gdn: >50m	7506	959
Needs Mods	-5782	2149
Dist CBD	1.92	0.49

RANDOM

Parameter	Coefficient	Standard Error
Community Level		
Constant	1553	451
ED Level		
Constant	487	67
Property Level		
Constant	1239	54

$-2*(\text{log-likelihood}) = -661.07$

Model 5.9 The full housing attributes model The full housing attribute model has the same random and property level fixed terms as the previous model, but now includes the locational attributes at the ED level (Table

5.12). Since these do not vary at level-1, the fixed and random estimates for the property level attributes remain unchanged. However, the ED level and community level random effects have been reduced, resulting in the property level explaining over a half of house price variations. The variable measuring housing quality is insignificant in the model (and hence was omitted), whilst social class has a significant effect upon house price differentials as is expected. These represent the relationship between house price and locational attributes after the compositional effects of the structural attributes have been allowed for. The addition of locational attributes at the ED level has resulted in marked changes at the community level. Firstly, the effects of area are now smaller, and on average, the premiums have halved for most communities. Furthermore, there has been some interesting changes in rank, notably the promotion of Riverside and Llandaff North. This suggests that these areas command a higher premium, given the social class of the areas, and may be caused by unaccounted externalities, in this case proximity to Bute Park and Llandaff Cathedral respectively. The cheapest communities are those characterised by a high proportion of old Local Authority stock and privately rented properties, implying a stigma effect. Again, the difference in the likelihood ratios (154.61) indicates that the addition of the locational attributes has had a significant effect upon explaining the variation in the model.

Higher Level Interactions

Introduction The above three models assume that the structural attributes are constant across Cardiff, and that the areal differences can be captured in a single variance term. An equivalent single level model would be similar to the traditional specification, except that the constant term would be expanded to accommodate community dummy variables. In such a specification, no structural parameter drift would occur, but instead there would be an additional premium to locate in a particular community. However such a specification would be inaccurate. Disequilibrium in supply and demand presuppose segmentation of the housing market, and this may lead to submarket formation, and hence the spatial variation of attribute prices. This was identified in the results from the spatial drift specifications, which indicated that the implicit prices of certain attributes appear to drift with respect to social class. In Model 5.9, a third of the house price variation occurs between communities after compositional effects of the housing stock and contextual effects of location have been taken into account. These unexplained community level differentials may

Table 5.12 Model 5.9 The full housing attributes model

FIXED

Parameter	Coefficient	Standard Error
Constant	44801	2321
Floor Area	36.11	1.69
Floor SD	-5.59	5.29
Floor D	3.56	0.28
Floor B	2.66	0.39
D Bath 2	5340	768
B Bath 2	-1532	615
D Shower 1	2076	445
Full CH	4092	924
Garage	5651	658
ORP	4244	804
Gdn: None	-4650	2114
Gdn: 5-50m	4443	859
Gdn: >50m	7506	959
Needs Mods	-5782	2149
Dist CBD	1.92	0.49
Social	3057	265.40
LA > 50%	-14798	3281
SocLA	-4635	907

RANDOM

Parameter	Coefficient	StandardError
Community Level		
Constant	702	209
ED Level		
Constant	255	50
Property Level		
Constant	1239	54

$-2*(\text{log-likelihood}) = -815.68$

be caused by variation in structural parameters at the community level. If the structural attributes are allowed to vary at the community level, any significant difference in the resulting random terms will be indicative of submarkets. Therefore, the random intercepts model (equation 5.7) was expanded with respect to equation 5.8:

$$\beta_{mjk} = \beta_m + \mu\beta_{mk} \qquad\qquad 5.8$$

To form the fully random model:

$$P_{ijk} = \alpha_{jk} X_{ijk} + \Sigma\, \beta_m Z_{mijk} +$$
$$(\mu\alpha_{jk}X_{ijk} + \mu\alpha_k X_{ijk} + \mu\beta_{mk} Z_{mijk} + \varepsilon_{ijk} X_{ijk}) \qquad\qquad 5.9$$

In which housing attribute Z_m is allowed to vary at the community level

Model 5.10 - Floor area interactions Since floor area is the principal structural attribute, this is allowed to vary at the community level. The random term in Table 5.13 measures the variation in the price of floor area between communities whilst the covariance term measures the relationship between average community level house price and the price of floor area. The addition of these random terms has caused a difference of 21.78, which is significant with 2 degrees of freedom. However, the standard error of the covariance term is rather large relative to the coefficient, suggesting that this term may have an insignificant effect upon explaining house price variation. This was verified by removing the covariance term, and re-estimating the model. There was negligible difference with the new parameter estimates whilst the new likelihood ratio (-864.58) indicated that the removal of the covariance term has had little effect upon the model. Hence, the model suggests that the price of floor area does indeed vary between communities, whilst the insignificance of the covariance term suggests that there is no relationship between average community level house price and the price of floor area. In other words, it is not the case that the increase in the price of floor area is greater for expensive properties than cheaper ones or vice versa.

Table 5.14 summarises the implicit prices of floor area for Cardiff and each community. On average across Cardiff, an extra square foot of space would be worth £38.34, whilst this varies from place to place. In Cyncoed, an extra square foot of space would be valued at £47.83, whilst in Adamsdown it would only be £26.84. These differences reflect the

Table 5.13 Model 5.10 The floor area - community interactions model

FIXED

Predictor	Coefficient	Standard Error
Constant	45705	1722
Floorr Area	39.11	3.94
Floor D	9.10	1.04
D Bath 2	4671	214
B Bath 2	-1532	615
D Shower 1	2778	154
Full CH	4524	854
Garage	5843	707
ORP	3510	880
Gdn: None	-4631	1759
Gdn: 5-50m	4050	844
Gdn: >50m	6790	917
Needs Mods	-5673	2033
Dist CBD	-0.9691	0.41
Social	3103	316
LA > 50%	- 15843	5835
SocLA	-4868	1160

RANDOM

Parameter	Coefficient	Standard Error
Community Level		
Constant	680	204
Floor Area	0.22	0.07
Foor Area / Constant	0.13	0.17
ED Level		
Constant	269	52
Property Level		
Constant	1266	57

-2*(log-likelihood) = -837.46

differences in the supply and demand schedules that operate in these communities. Figure 5.8 shows the change that the addition of the random floor area term has had on the implicit price of community. It is clear that there is an Inner Area/suburban split, with suburban communities becoming more expensive once the differential price of floor area has been taken into account and vice versa. However, the size of this effect is small, and there has been little change in rank.

The inclusion of the floor area random term has had a small, but important effect upon the attribute prices at the individual property level. Firstly, the variable measuring bungalow floor area has became insignificant. This implies this was probably capturing the differential floor area price now accounted for by the higher level random terms. More specifically, it would appear that the variable was capturing the contextual effects operating upon floor area in Cyncoed and Rhiwbina, the location of the majority of bungalows. Secondly, the measure of accessibility to the CBD has become negative and significant. Again, this would imply that the variable had captured the contextual effects of community upon floor area. Table 5.14 illustrates that floor area in suburban communities are generally more expensive than Inner Area ones, and hence the counter-intuitive positive relationship between distance and house price. The price of floor area for detached properties does not vary significantly between communities, but is uniformly more expensive than all other properties across the city as a whole. The remaining structural attributes are unchanged, and none of them varied significantly at the community level.

Model 5.11 Social class - between community level interactions Previous empirical research has suggested that social class is an important factor in house price variation, and this was tested using the spatial drift specification, which established that the implicit price of certain attributes, such as floor area, interacted with social class. Model 5.11 captures this effect by allowing social-economic class to vary at the community level. This model also allows the affect of social class to vary depending upon community context, a concept that was not possible in the spatial drift specification, which assumed a constant effect. Table 5.15 shows that, in terms of the fixed attribute prices, the counter-intuitive price differential between single and two bathroom bungalows becomes insignificant and falls out of the model, indicating that it was capturing a social class effect.

The significance of the three additional random terms were evaluated by the size of their standard errors, and the likelihood ratio statistic. These indicated that the random term for social class has had an

Table 5.14 Community level floor area differential prices

Model 5.10	Price	Model 5.11	Price	Model 5.12	Price
Cyncoed	9.49	Llanishen	9.20	Llanishen	9.11
Llanishen	8.71	Cyncoed	9.16	Cyncoed	8.88
Whitchurch & Tongwynlais	6.27	Whitchurch & Tongwynlais	6.52	Whitchurch & Tongwynlais	6.24
Llandaff	4.37	Llandaff	4.78	Llandaff	4.47
Roath	3.95	Roath	4.39	Roath	4.26
Rhiwbina	3.21	Rhiwbina	3.50	Rhiwbina	3.70
Plasnewydd	2.92	Llanrumney	3.14	Plasnewydd	3.03
Llanrumney	2.43	Plasnewydd	3.04	Llanrumney	2.74
Heath	1.90	Heath	2.53	Heath	2.07
Caerau	0.48	Fairwater	0.84	Fairwater	0.80
Cathays	0.44	Lisvane & St Mellons	0.43	Lisvane & St Mellons	0.62
Lisvane & St Mellons	0.06	Radyr & St Fagans	0.13	Radyr & St Fagans	0.15
Fairwater	0.04	Caerau	0.07	Riverside	-0.12
Butetown	-0.01	Llandaff North	-0.07	Caerau	-0.14
Trowbridge	-0.30	Riverside	-0.14	Cathays	-0.14
Radyr & St Fagans	-0.34	Ely	-0.47	Llandaff North	-0.23
Llandaff North	-0.53	Cathays	-1.15	Butetown	-0.37
Riverside	-0.63	Pentwyn	-1.25	Ely	-1.48
Ely	-1.39	Trowbridge	-2.17	Pentwyn	-1.80
Pentwyn	-1.51	Gabalfa	-3.02	Trowbridge	-1.96
Gabalfa	-3.01	Canton	-3.19	Gabalfa	-2.78
Rumney	-3.18	Rumney	-3.27	Canton	-3.08
Canton	-3.75	Butetown	-3.60	Rumney	-3.27
Grangetown	-7.25	Grangetown	-6.44	Grangetown	-7.53
Splott	-10.80	Splott	-10.8	Splott	-11.40
Adamsdown	-11.50	Adamsdown	-12.10	Adamsdown	-11.80

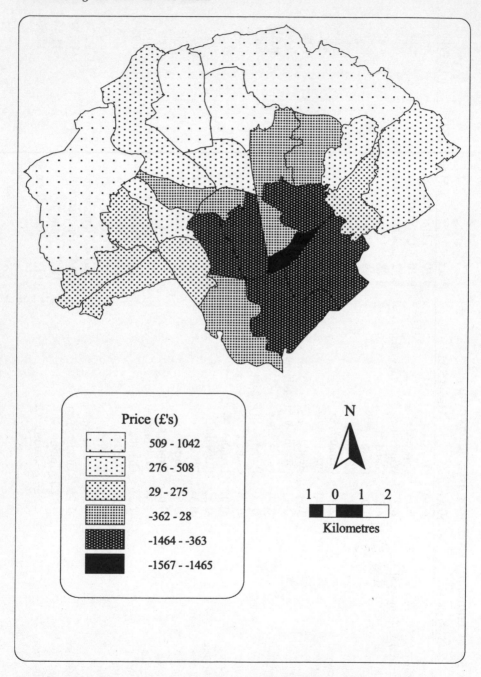

**Figure 5.8 Changes in community level prices due to floor area
interactions**

Table 5.15 Model 5.11 Social class - between community level interactions model

FIXED

Predictor	Coefficient	Standard Error
Constant	45252	1639
Floor Area	38.34	3.955
Floor D	8.57	1.031
D Bath 2	4511	2136
D Shower 1	2763	1529
Full CH	4414	831
Garage	5734	727
ORP	3470	827
Gdn: None	-4955	1772
Gdn: 5-50m	3843	834
Gdn: >50m	6556	899
Needs Mods	-5556	2069
Dist CBD	-0.7021	0.422
Social	2983	368
LA > 50%	-10161	3461
SocLA	-3644	1025

RANDOM

Parameter	Coefficient	Standard Error
Community Level		
Constant	633	198
Floor Area	0.00089	0.00039
Social	30.65	15.88
Floor Area / Constant	0.25	0.083
Floor Area / Social	0.053	0.019
Constant / Social	15.79	39.47
ED Level		
Constant	215	48
Property Level		
Constant	1246	56

-2*(log-likelihood) = -865.76

insignificant effect upon the model, whilst the floor area / constant covariance term, which was previously insignificant, now had a significant effect. Therefore, although the affect of social class *per se* does not vary significantly between communities, it does interact with floor area and average house price at this higher level, such that the marginal price / floor area relationship is steeper in areas of higher social class, whilst average house prices are more expensive. As a consequence, the covariance between floor area and typical community level house price has now become significant, with Figure 5.9 implying that the differences in community level property prices are generally smaller for larger houses than smaller houses. These results suggest that the difference a community makes depends upon the size of the property, and the social class of the area. An examination of the community effects (Table 5.10) indicates that the most expensive communities are those with a combination of large properties and high social class, and vice-versa. Areas of high social class and (relatively) small properties, such as Lisvane and St Mellons, or large properties and average social class, such as Radyr & St Fagans and Rhiwbina, only have average premiums.

Model 5.12 Social class - within community level interactions The above model has allowed social class to vary at the community level. However, the spatial parameter drift models showed that social class operates in a non-linear fashion, such that the greatest effects occur in areas of either very high or very low social class. This was specified through the use of a quadratic functional form. In the multi-level specification, however, such an effect can be achieved by allowing social class to vary within communities, at the level of the ED. Coupled with variations at the community level, such a model will allow a complex geography of social class interactions. Table 5.16 summarises the results of this model. The influence of these ED level random terms are substantial (differences in likelihood of 18.12 with 2 degrees of freedom). The constant term becomes insignificant, implying that within a community, average house price does not vary significantly at the ED level, whilst the significant social class random term accounts for the non-linear affect of social class.

At the community level, the significance of the random terms become greater (since their standard errors decrease), although the random term for social class still remains insignificant. An examination of the changes in the floor area differentials suggests that those communities with the greatest mix of social class have experienced the greatest changes. For instance, Butetown, which is traditionally a low social class area but now

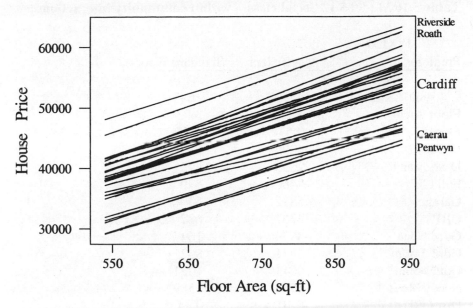

Figure 5.9 Price / floor area relationship with social class interactions: model 5.11

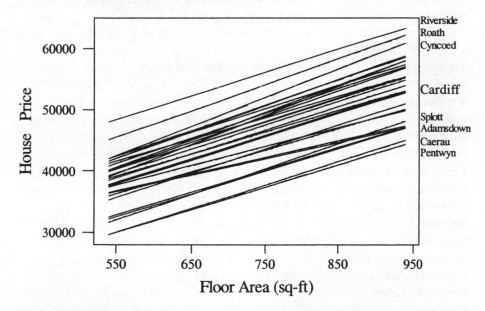

Figure 5.10 Price / floor area relationship with social class interactions: model 5.12

Table 5.16 Model 5.12 Social class - within community interactions

FIXED		
Predictor	Coefficient	Standard Error
Constant	45707	1645
Floor Area	38.62	3.93
Floor D	9.53	0.75
D Bath 2	4206	619
D Shower 1	2553	492
Full CH	4460	779
Garage	5822	639
ORP	3267	687
Gdn: None	-4974	1647
Gdn: 5-50m	3921	862
Gdn: >50m	6643	1021
Needs Mods	-5529	1589
Dist CBD	-0.10	0.34
Social	3333	283
LA > 50%	-11350	4687
SocLA	-3606	1347

RANDOM		
Parameter	Coefficient	Standard Error
Community Level		
Constant	588	177
Floor Area	0.00087	0.00023
Social	14.17	11.24
Floor Area / Constant	0.27	0.084
Floor Area/ Social	0.025	0.0088
Constant / Social	21.32	7.84
ED Level		
Constant	90	51
Social	44.27	15.70
Constant / Social	7.30	15.22
Property Level		
Constant	1245	56

-2*(log-likelihood) = -883.88

has enclaves of high social class in the redeveloped docklands, has experienced an increase in floor area of £3.23 per square foot. Conversely, in neighbouring Grangetown, the floor area differential has declined in price by £1.10 per square foot, reflecting the higher concentrations of lower class areas. Figure 5.10 describes the relationship between average community level house price and the price of floor area. The effect of allowing social class to vary within a community has increased the strength of this relationship in the more expensive communities, such as Cyncoed, whilst having very little influence on communities of below average price. Two communities that appear to have a significantly different relationship from the rest of Cardiff are Splott and Adamsdown, which have much gentler price gradients. Since these are adjacent communities, the supply and demand mechanisms in these communities maybe very similar, compared to the rest of Cardiff.

Figure 5.11 illustrates the geography of the differential community prices. The most expensive communities tend to be located either in, or adjacent to the Inner Area, which contrasts with Figure 5.7, in which the most expensive were on the edge of the suburbs. The cheapest communities correspond to the peripheral Local Authority estates, corroborating the evidence of the 'stigma effect' attached to these areas. This gives an idea of which communities require additional premiums, and could be viewed as a measure of community desirability.

The Relationship Between Social Class and Floor Area

It was previously argued that in areas of higher social class, buyers spend marginally more on locational attributes than on structural attributes, and this is reflected in the decreasing price per unit floor area. This assertion can be re-evaluated using the results from the multi-level models. Figure 5.12 is the result of plotting the price of unit floor area in each community (Model 5.12) against community level social class. With the exception of Butetown, it can be seen that the price of floor area initially increases, before decreasing in areas of high social class. Butetown is an anomaly, caused by the Docklands development in an area traditionally of low social class.

Figure 5.13 describes the relationship between price per unit floor area in a community and average community level house price, both deviated around their means. Hence the centre of the graph is the Cardiff average. This summarises the previous finding of a positive linear relationship between the price of floor area and the average value of a

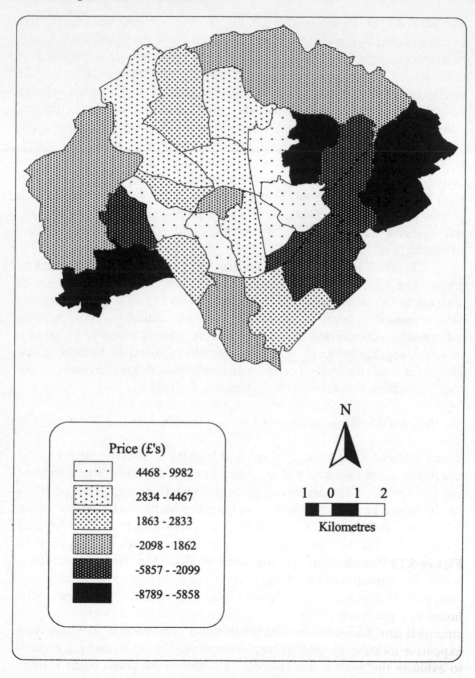

Figure 5.11 Differential community level prices: model 5.12

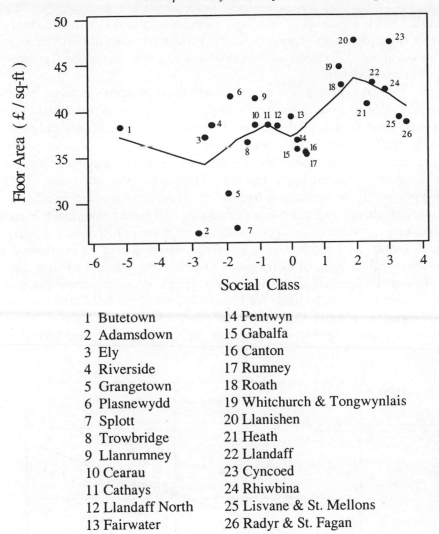

Figure 5.12 Price per unit floor area by community level social class: model 5.12

1 Butetown	14 Pentwyn
2 Adamsdown	15 Gabalfa
3 Ely	16 Canton
4 Riverside	17 Rumney
5 Grangetown	18 Roath
6 Plasnewydd	19 Whitchurch & Tongwynlais
7 Splott	20 Llanishen
8 Trowbridge	21 Heath
9 Llanrumney	22 Llandaff
10 Cearau	23 Cyncoed
11 Cathays	24 Rhiwbina
12 Llandaff North	25 Lisvane & St. Mellons
13 Fairwater	26 Radyr & St. Fagan

house in a particular community. If it is assumed that the most expensive structural attributes will be located in those communities with the most expensive locational attributes, as the graph predicts, then this can be used to evaluate the relationship between the value of housing in terms of its structural and locational attributes. For instance, in Roath, Cyncoed and Rhiwbina, buyers are spending marginally more on both the physical housing stock and location than buyers in Splott and Adamsdown. Hence, it can be expected that residents in the former communities will be more anxious about the effects of negative externalities upon their property prices than the latter.

A more interesting, and less trivial situation occurs in communities that do not fit into this general relationship. These tend to be located in the top, left-hand section of the graph (Riverside, Cathays, Canton, Llandaff North and Butetown) and the bottom right-hand side (Fairwater and Llanrumney). In the case of the former communities, marginally more is spent upon locational attributes than upon structural attributes of the property whilst the opposite is true for the latter communities. In all these cases, any change in locational externalities will have a marginally bigger affect upon property prices. For instance, in the case of Riverside, Cathays and Canton, the proximity of Bute Park would appear to be having a beneficial affect upon property prices. Hence, any changes to Bute Park may have a large negative effect upon prices in these communities. A similar argument can be made for Llandaff North (Llandaff Cathedral) and Butetown (Docklands). Conversely, property prices in the communities of Fairwater and Llanrumney may be being depressed by locational

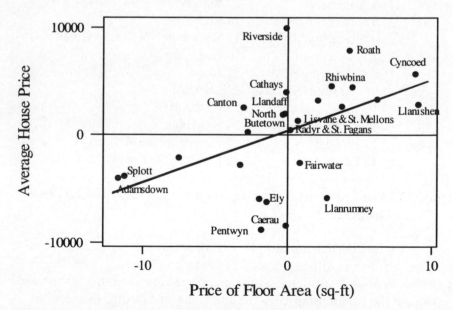

Figure 5.13 Relationship between the differential price of floor area and average community level house prices

externalities, and the improvement of these may increase property prices disproportionately compared to other communities. Finally, it is worth noting that the communities of Lisvane & St. Mellons and Radyr & St. Fagans correspond to the Cardiff average with respect to their structural attributes and community premium prices. Hence, buyers in these communities are getting value for money in terms of house size and location, even though house prices in these communities are the highest in the whole city.

Figure 5.14 describes the value of housing in Local Authority areas once community context has been taken into account. Once again, houses located in areas in which the majority of properties are Local Authority owned are more expensive *ceteris paribus*, compared to other areas of low social class. However, this premium has declined in absolute value, whilst the 'stigma effect' now occurs in areas of much lower socio-economic class (in this case, in areas with scores greater than -5). This suggests that the spatial parameter drift models were under-estimating this 'stigma effect'. An examination of Figure 5.13 shows that the communities in which the majority of Local Authority areas are located (Ely, Caerau, and Pentwyn) have cheaper premiums than expected given the price of floor area, and hence it is the failure of the spatial drift specification to take into

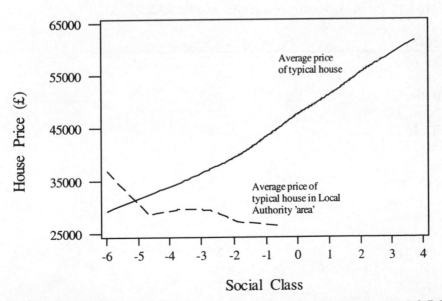

Figure 5.14 Local Authority housing stock and social class: model 5.12

account these community effects that leads to this under-estimation. Figure 5.14 also shows how the multi-level specification allows the influence of social class to vary in a non-linear fashion, with the 'stigma-effect' initially decreasing with social class, and then increasing again.

Tests for heteroscedasticity and spatial autocorrelation were undertaken to evaluate the degree to which the final model, Model 5.12, captures the spatial structures in the data. Table 5.17 is a summary of Breusch Pagan test statistics for each variable in the fixed part of the model. These suggest that heteroscedasticity is still present in the floor area and garden size attributes, although the effects are markedly reduced relative to those of Model 5.6. This heteroscedasticity is attributable either to variance heteroscedasticity in the attribute variables concerned, or the continued presence of spatial heteroscedasticity. The former would occur if higher income buyers spend marginally less on structural attributes and more on locational attributes than low income buyers. In such a case, error variance associated with high income buyers would be different than for low income buyers. Spatial heteroscedasticity is likely to occur since it is unlikely that the communities correspond perfectly with submarkets, and hence communities will not be totally homogeneous with respect to supply

Table 5.17 Breusch-Pagan results for model 5.12

Variable	Breusch Pagan
Floor Area	21.36
Floor D	5.2
D Baths 2	0.003
D Shower 1	0.72
Full CH	0.58
Garage	4.6
ORP	1.02
Gdn:None	12.7
Gdn:5-50m	12.9
Gdn:>50m	0.9
Needs Mod	0.16
Dist CBD	0.87
Social	0.09
LA > 50%	3.7
SocLA	3.9

and demand schedules. This will lead to structural instability of parameter estimates. With regards to spatial autocorrelation, the Moran test confirms that spatial autocorrelation is no longer significant ($I = 0.00067$). This is to be expected however, since the multi-level specification will model spatial autocorrelation regardless of whether submarkets are correctly delimited. Although the specification still needs slight improvements, it would appear that it has successfully captured the spatial structures in the data and therefore the spatial dynamics of the Cardiff housing market.

A Comparison of the Specifications

It is clear that each specification conceptualises the Cardiff housing market as operating in a slightly different way. The ability of each to model the spatial dynamics of the housing market has relied upon theoretical considerations of how the housing market should operate, and diagnostic tests, to check how each model accounts for the spatial elements of the data. Taken together, these considerations suggest that the multi-level specification is the most suitable. However, this does not necessarily mean that the other specifications incorrectly estimate implicit prices. Rather, it may be the case that some housing attributes are robust to changes in specification. To ascertain the robustness of the models, the parameters of three of the models were compared - Model 5.2, Model 5.6 Model 5.12. These are the most sophisticated models for each of the specifications. The first thing to note is that, with the exception of floor area, Model 5.2 and Model 5.6 produce the most comparable parameter estimates. The estimates of Model 5.12 tend to be smaller. It is interesting to note the differences in the detached house variables. These are much smaller in the multi-level model, implying that these variables had captured locational differences in the previous models. Only the multi-level specification was able to model compositional and contextual effects simultaneously. There are several variables, such as garden size, which appear to be robust, and do not vary to a significant degree between the models. These tend to be structural attributes external to the physical dwelling, and this suggests that they may be valued differently to internal structural attributes.

Estimated rent gradient for model 5.1

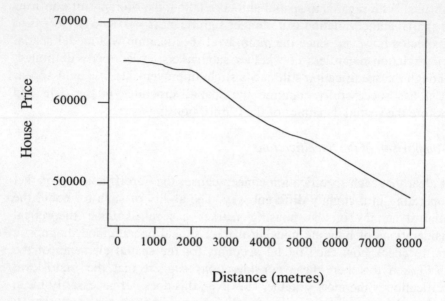

Estimate rent gradient for model 5.2

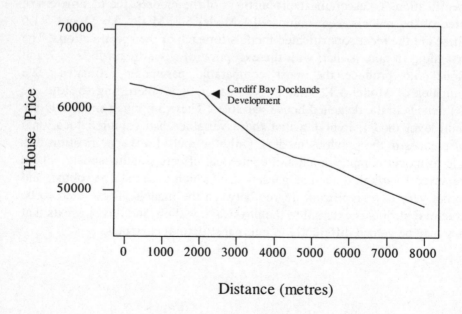

Figure 5.15 Estimated rent gradients for selected models

Estimated rent gradient for model 5.6

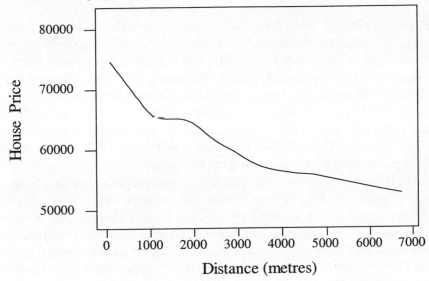

Estimated rent gradient for model 5.12

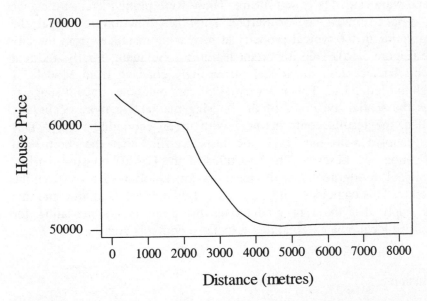

Figure 5.15 (cont) Estimated rent gradients for selected models

The Cardiff Rent Gradient

The rent gradient is one of the basic tenets of standard urban economic theory, and hence its estimation is an important result, particularly in light of past research. A linear functional form was used to estimate the rent gradient in all of the previous models. The resulting gradients ranged from £0.70 to £1.90 per metre, suggesting an average decline in property prices of around £130 per kilometre. Although this is quite a low value, Cardiff is a comparatively small city, and has relatively good transport routes from the city centre to the suburbs. This may result in accessibility not being valued particularly highly compared to other cities.

However a linear rent gradient is counter-intuitive, and a non-linear gradient, such as inverse power and negative exponential may be more appropriate, as postulated by urban economic theory. The problem lies in the fact that, despite what previous researchers have written, the exact functional form of the rent gradient cannot be known *a priori* . Also there is no reason why the rent gradient should approximate any standard functional form. Thus, to avoid constraining the model unnecessarily, four of the models were re-estimated using the dummy distance interval measures in Table 4.6. The estimated implicit prices all had the correct sign and were statistically significant. These were plotted against distance using a locally weighted scatterplot smoother function to reveal the average price of the typical property at increasing distances from the city centre (Figure 5.15). Two important features are evident. Firstly, the rent gradients become less linear and increasingly concave from Model 5.1 through to Model 5.12. This is a very significant outcome, since it suggests that as the spatial dynamics of the housing market are more accurately modelled, the results conform more with urban economic theory. This result, coupled with the diagnostic tests, verifies that the multi-level specification best describes the structures of the Cardiff housing market. Secondly, all four graphs reveal a local maxima at around 2.5 km from the city centre. This coincides with the Cardiff Bay development area and this would imply that the rent gradient at this point is compensating for unspecified locational externalities associated with this area.

Conclusion

The general aim of this chapter has been to model the spatial dynamics of the Cardiff Housing market. In doing so, several substantive results have been established. Firstly, on the technical and theoretical side, the three different specifications have been examined, and their ability to model spatial data evaluated. It has been shown that the ability to capture spatial effects has substantial influence upon the results. In particular, failure to take into account how structural attributes vary with locational context can lead to heteroscedasticity and spatial autocorrelation in the residuals, and incorrect implicit prices, such as the negative result for bungalows with two bathrooms. Moreover, the advantages of modelling space using a multi-level specification as opposed to a single level one has also been demonstrated. The multi-level specification allows both compositional and contextual effects to be modelled simultaneously, allowing location attributes, such as social class, to vary with context.

The results also demonstrated the existence of submarkets, delimited according to both the housing stock and by geographical area. By using the multi-level specification, the price of floor area - the most important structural attribute - was seen to vary significantly between communities and housing type. The final model concluded that the price of floor area depended upon both the average price of a community and the social class of the area, and that separate market conditions exist for detached houses, irrespective of location.

The results also confirmed the importance of locational attributes in determining house prices. Social class was consistently the second most important variable in the model, and was highly significant. When the structural attributes were allowed to vary with social class, it was demonstrated that the price of floor area in areas of high social class were marginally less than in areas of average social class, implying that marginally more was spent upon locational attributes. The importance of locational externalities upon house prices are examined in more detail in the next chapter, in which the influence of specific locational attributes shall be modelled for properties in the Inner Area of Cardiff.

6 Towards a Valuation of Locational Externalities

Introduction

The previous chapter has explored the spatial dynamics of a typical UK housing market. This chapter aims to expand this by evaluating the influence of specific locational externalities using the multi-level specification. Four levels of resolution have been defined for this purpose: the property level, the (sub)-street level, the HCS Area level and the community level. Table 6.1 is a summary of the locational attributes that are hypothesized to influence house prices in the Inner Area. These have been grouped according to the spatial level at which they are conceptualised to operate. Social class has now been operationalised at the community level, since it is no longer acting as a surrogate variable for locational attributes, but as a contextual background against which supply and demand mechanisms operate. In contrast, the variable measuring areas of predominately Local Authority tenure was aggregated to the HCS Area Level, reflecting the very localised nature of this tenure within the Inner Area.

The Spatially Invariant Housing Market

Introduction

This is based upon the random intercept model, that allows the average price of property (i.e. the constant term) to vary at the different levels of resolution, but keeps the implicit prices of the housing attributes constant. The corollary of this model is an assumption of a spatially homogeneous housing market, where supply and demand schedules are uniform and no submarkets exist.

Table 6.1 Inner Area locational attributes

Attribute	Abbrev.	Attribute	Abbrev.
Accessibility to CBD	Dist CBD	Street quality 0-50m: Below Ave	Below Ave 0-50m
Accessibility to M4 motorway	Dist MWAY	Street quality 0-50m: Above Ave	Above Ave 0-50m
Accessibility to railway stations	Dist Station	Street quality 0-50m: Good	Good 0-50m
Proximity to hospitals	Hospitals	Street quality 50-100m: Poor	Poor 0-50m
Proximity to sports centres	Sports	Street quality 50-100m: Below Ave	Below Ave 50-100m
Proximity to community centres	Community	Street quality 50-100m: Above Ave	Above Ave 50-100m
Proximity to institutional centres	Institutional	Street quality 50-100m: Good	Good 50-100m
Proximity to local shops	Shops	Street quality 100-200m: Poor	Poor 100-200m
Proximity to primary schools	Primary	Street quality 100-200m: Below Ave	Below Ave 100-200m
Proximity to secondary schools	Secondary	Street quality 100-200m: Above Ave	Above Ave 100-200m
Proximity to Bute Park	Bute Park	Street quality 100-200m: Good	Good 100-200m
Proximity to parks / open space	Parks	Street non-residential landuse.	Non-Res Buildings
Proximity to light industrial landuse	Light Ind	School Catchment: Willows High	Sch: Willows
Proximity to heavy industrial landuse	Heavy Ind	School Catchment: Fitzalan High	Sch: Fitzalan
Rail 0 -50m	Rail 0 -50m	School Catchment: Cantonia High	Sch: Cantonia
Rail 50 - 100m	Rail 50 - 100m	School Catchment: Cathays High	Sch: Cathays
Rail 100 - 150m	Rail 100 - 150m	School Catchment: St Teilo's High	Sch: St Teilo's
Rail 150 - 200m	Rail 150 - 200m	% Local Authority tenure	LA > 50%
River 0 - 50m	River 0 - 50m	% open space	%Open Space
River 50 - 100m	River 50 - 100m	% non-residential landuse	% Non-Residential
River 100 - 150m	River 100 - 150m	Housing density	Density
River 150 - 200m	River 150 - 200m	Quality of local shops	Q.Shop
Road Type: Primary	Primary Road	Quality of local public transport	Q.Transport
Road Type: Secondary	Secondary Road	Quality of local sport facilities	Q. Sport
Road Type: Residential	Residential Road	Quality of local parks	Q.Parks
Road Type: Cul-de-sac / Close	Close	Quality of community facilities	Q.Community
Street quality 0-50m: Poor	Poor 0-50m	Social economic class	Social

Model 6.1 The Grand Mean Model

Table 6.2 summarises this basic model, which estimates an average house price for the whole of the Inner Area of around £53,500. The variation around this average price has been decomposed into variations at the level of the property, the street, the HCS Area and the community. With respect to each level, the greatest proportion of the total variation in house price occurs between communities (£6,828), with the least between HCS Areas (£2,033) within a community.

Table 6.2 Model 6.1 The grand mean model

FIXED

Predictor	Coefficient	Standard Error
Constant	53637	423.68

RANDOM

Parameter	Coefficient	Standard Error
Community Level		
Constant	6829	3129.83
HCS Area Level		
Constant	2033	637.61
Street Level		
Constant	3640	672.61
Property Level		
Constant	4903	493.07

-2*(log-likelihood) = 317.04

Model 6.2 The Structural Attributes Model

To isolate the influence of locational externalities, the structural attributes were first added to the grand means model to account for the compositional effects of the housing stock (Table 6.3). The constant term now represents the price of an averaged sized terraced property (£43,477). The most

Table 6.3 Model 6.2 The structural attributes model

FIXED

Predictor	Coefficient	Standard Error
Constant	43477	218.74
Floor Area	31.18	1.43
Floor D	8.66	2.97
MT Bath 2	9807	2987.88
Full CH	3751	830.39
Garage	3933	883.78
ORP	4173	1008.99
Gdn: None	-4334	1111.06
Gdn: 5-50m	2657	972.46
Needs Mods	-3698	1091.62

RANDOM

Parameter	Coefficient	Standard Error
Community Level		
Constant	1461	723.20
HCS Area Level		
Constant	910	238.63
Street Level		
Constant	830	196.81
Property Level		
Constant	1620	163.79

-2*(log-likelihood) = -339.123

influential variable is again floor area, with the model also revealing the separate market conditions for detached housing. The variable measuring shower rooms was insignificant, whilst the value of modernising a property was estimated to be around £3,700.

The inclusion of the structural attributes has resulted in a large decline in the variance at all levels, but particularly at the community level, suggesting that the differences between communities in Model 6.1 was caused principally by differences in the housing stock. This has resulted in the property level and the community level variables explaining roughly the same amount of variation in house price. Under the null hypothesis, the difference of the likelihood's (656.16) follows a chi-square distribution with 9 degrees of freedom. The probability of obtaining a chi-square of this magnitude by chance is negligible (less than 0.001), strongly indicating that the structural attributes have an important effect in explaining house price variation in the model.

Model 6.3: The Property Level Locational Attributes Model

Introduction The previous model has taken into account the compositional effects of the housing stock. It is now time to evaluate the affects of the locational attributes at each of the four levels, starting with the level of the individual property. The locational attributes that operate at this level are those associated with accessibility to work place and proximity to non-residential landuses.

Accessibility Three measures of accessibility were modelled: access to the CBD, access to the nearest motorway junction and access to the nearest railway station. A linear functional form was used for accessibility to the CBD and the motorway, and a negative exponential was used for accessibility to the railway stations, since it is hypothesized that this would only be significant over very short distances. Table 6.4 is a summary of the estimated parameters. It can be seen that only accessibility to the nearest motorway junction is significant. The insignificance of the remaining two can be explained by the fact that the majority of properties are within close enough proximity to make accessibility to the CBD and railway stations relatively unimportant. However, it is interesting to note the similarity between the parameter estimate for accessibility to the motorway, and that for accessibility to the CBD estimated in the previous chapter.

Table 6.4 Accessibility parameter estimates for model 6.3

Predictor	Coefficient	Standard Error
Dist CBD	-1.52	1.53
Dist MWAY	-2.65	1.21
Dist Station($\beta = -2.0$)	4.06	3.32

Proximity to non-residential landuse Table 6.5 is a summary of the parameter estimates of proximity to railway lines and water-fronts, measured by the use of dummy distance intervals described in Chapter four.

Table 6.5 Railway lines and River Taff parameter estimates

Predictor	Coefficient	Standard Error
Rail 0- 50m	-2790	1401.55
Rail 50-100m	-86	1343.09
Rail 50-150m	-10645	1288.78
Rail 150-200m	-793	1292.45
River 0-50m	8866	2747.14
River 50-100m	-4912	2665.45
River 100-150m	-1233	1878.92
River 150 200m	-1891	1698.70

It can be seen that, for both railway lines and rivers, only properties within fifty metres are significantly affected. Since these represent properties directly facing them, it can be regarded as an aesthetic cost. The model suggests that railway lines reduce the value of a property by around £2,800, whilst close proximity to river will increase the value of a property by nearly £9,000. This is a very high value and may be a surrogate for lower densities and / or access to openspace given the propensity of the River Taff to flood. As such, these values shall be re-evaluated later.

The second measure of proximity was based upon continuous distance from the externality, taking into account the underlying topology. These measures incorporate an attractiveness index to model the magnitude of the externality, and a distance decay function to model proximity (see equation 4.1). For areas of non-residential landuse, such as industrial sites and parkland, the magnitude was calculated as the area of land squared. For other non-residential landuses, such as shops and schools, the attractiveness index was set to unity. Since the shape of the distance decay function was not known *a priori*, five β-values, ranging from 0.25 - 3.0, were used to calibrate five distance decay functions. A small β-value represents a gentle distance decay curve and hence the greater the extent of the effect. A large β-value represents a steep distance decay curve, and the externality only has an influence over a short distance. The aim of the research is not to find the exact β-value for each externality *per se*, but to discover over what range of β-values the externality effect is significant, and then to compare the relative affects of each externality. Thus, each of the five β-values was modelled separately, and the significance of the parameter used to determine the appropriateness of the β-value.

Table 6.6 is a summary of the t-statistics of the estimated externalities. An insignificant result for all values suggests that the externality has a very steep distance decay function and hence has a negligible affect upon property prices. Conversely, the larger the t-statistic, the better the β-value captures the affect of the externality. For instance, it can be seen that the affect of Bute Park is significant between the range of 0.25-2.0. However, the t-statistics decrease in magnitude (and hence significance) as the β-values increase. This suggests that Bute Park has a gentle distance decay function, best approximated by a small β-value.

The insignificance of the t-statistics for the overall affects of non-residential landuse confirms that property prices are not influenced by non-residential landuse *per se*, but by specific non-residential landuses. For this

Table 6.6 T-statistics for non-residential landuse proximity estimates

β-values	0.25	0.5	1.0	2.0	3.0
Non-residential Landuse	0.37	0.09	0.36	0.96	1.74
Bute Park	3.86	3.60	2.93	2.00	1.69
Parks	3.32	2.78	2.24	1.06	0.68
Industrial	0.02	0.05	0.05	1.11	0.80
Industrial: Heavy	2.01	2.27	1.51	0.48	0.11
Industrial: Light	0.30	0.47	0.28	1.03	0.76
Community	0.90	1.59	1.97	1.94	1.92
Institutional	0.22	0.668	0.23	0.13	0.01
Hospital	1.18	0.98	0.61	0.26	0.29
Sports	0.88	0.016	1.21	2.09	2.06
Shops	0.29	0.27	0.02	0.121	0.01
Primary School	1.83	1.76	1.77	2.03	2.14
Secondary School	1.66	1.56	1.30	0.88	0.69

reason non-residential landuses were separated into various classifications of landuses, the primary classifications being industrial sites and open space. The t-statistics for parks follow a similar pattern to Bute Park, although they are smaller in magnitude, suggesting less of an impact. Industrial landuse has an insignificant impact upon property prices until the distinction is made between 'light' industrial areas and 'heavy' industrial areas. Heavy industrial areas have the most significant effect with a distance decay β-value of around 0.5, becoming insignificant at values greater or less than this. This implies that although the spatial affects of heavy industrial areas are quite extensive, they are not as extensive as parks or open space. Hence, it may be the case that buyers and sellers of property emphasize proximity to positive externalities, such as parks and downplay proximity to negative externalities, such as heavy industrial areas. The t-statistics suggest that proximity to light industrial areas have a negligible affect on property prices, although this may be a result of the fact that the sample contained very few properties significantly near these areas.

The remaining externalities were estimated with an attractiveness index set to unity and hence the externality effect is solely determined by proximity. Proximity to community centres has a β-value of around 1.0, at which it is marginally significant. This implies that the externality effect is directly proportional to distance. Institutional centres are insignificant for the whole range of β-values, suggesting that they have very little impact of property prices. The externality effects of hospitals are also insignificant for the range of β-values. However, there is a trend of the t-statistics increasing with decreasing β-values, indicating that the externality effect may become significant for very small values of β. This hypothesis was tested using β-values within the range of 0.125 - 0.0156, which produced parameters with t-statistics of between 1.71 - 1.89, bordering upon significance. This implies that hospitals have a very gentle distance decay curve, influencing property prices across a much wider area than that being studied. If this is the case, then the Inner Area will be too small to capture this effect.

The results for proximity to sport centres implies that their externality effects operate across small distances, with its optimal β-value falling within the range of 2.0 - 3.0. This suggests that sport centres are only influential if they are within walking distance. A similar result applies to proximity to primary schools, but not secondary schools. This is an interesting result since it suggests that walking distance to schools is only a

Table 6.7 Model 6.3 Property level locational attributes

FIXED

Predictor	Coefficient	Standard Error
Constant	43537	222.31
Floor Area	31.01	1.39
Floor D	8.71	2.86
MT Bath 2	7524	2811.18
Full CH	4000	797.75
Garage	3947	849.61
ORP	3662	963.10
Gdn: None	-4885	1080.35
Gdn: 5-50m	2255	924.37
Needs Mods	-3656	1051.65
Dist MWAY	-2.65	1.21
Bute Park($\beta = 0.25$)	11644	3016.68
Parks($\beta = 0.25$)	159256	47997.53
Heavy Ind($\beta = 0.5$)	-24960	10979.76
Sports($\beta = 2.0$)	0.0002236	0.0001070
Primary($\beta = 3$)	0.0004350	0.0002030
Rail 0-50m	-2790	1401.55
River 0-50m	8869	2747.14

RANDOM

Parameter	Coefficent	Standard Error
Community Level		
Constant	944	469.54
HCS Area Level		
Constant	468	153.68
Street Level		
Constant	718	177.75
Property Level		
Constant	1503	150.90

-2*(log-likelihood) = -406.374

consideration for households with small children. However, the catchment area of a secondary school may be important, and this is evaluated later. Finally, the insignificant results for shops can be explained by the close proximity to the city centre and the high density of smaller shopping areas in the Inner Area.

The externality parameter estimates Table 6.7 presents Model 6.3 with the significant property level locational attributes. The chi-squared test indicates that the difference of the likelihood statistics (67.25) is significant at the 99% level with 8 degrees of freedom, implying that the property level locational attributes have had an important statistical effect in explaining house price variation. The random terms in the model suggests that the addition of the property level locational attributes has had the greatest effect of explaining higher level variation, particularly at the HCS Area and community level, with only a negligible effect upon the variation between properties. This is understandable since at the property level, the majority of variation is caused by differences in structural attributes, whilst locational attributes would tend to be quite similar at this level.

The magnitude and geography of property level externalities Looking at the results of Model 6.3 in more detail, it is possible to arrive at some interesting conclusions concerning the affects of positive and negative externalities and the interaction between them. The coefficients for Parks and Heavy Industry have been standardised in order that they show the influence of one kilometre squared of the landuse, at one kilometre distance from a property. For instance, the coefficient for Heavy Industry is -24960. This means that one kilometre square of heavy industrial area will decrease the price of a property located one kilometre away by £24,960. A similar explanation applies to the Parks coefficient. However, it would be unwise to extrapolate too much beyond the range of values estimated for each landuse in Cardiff. For heavy industrial sites, this is within the range of 0.165 - 0.358 kilometres squared in area, whilst for parks, the range falls within 0.0062 - 0.130 kilometres squared in area. For a property located one kilometre away from these landuses, the estimated effects on price for each of these ranges would be £1,150 - £4,260 and £6 - £2,691 respectively. Imposing such parameters will prevent excessive predictions being made using the model. Figures 6.1 & 6.2 illustrate the estimated distance decay curves for heavy industrial sites and parks and open space. It can be seen how the externality effects decrease with increasing distance

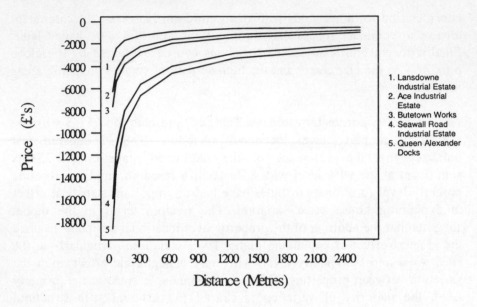

Figure 6.1 Model 6.3 Valuing proximity to heavy industrial areas

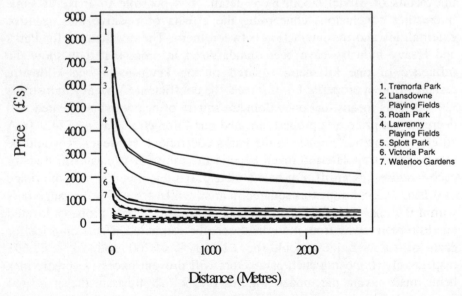

Figure 6.2 Model 6.3 Valuing proximity to parks and open space

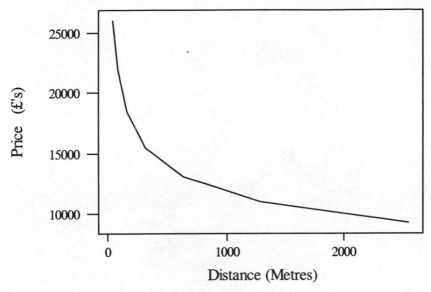

Figure 6.3 Model 6.3 Valuing proximity to Bute Park

from the source, and how they tend towards convergence. Both externalities have the greatest impact up to 0.25 kilometres from the landuse, suggesting that visible presence may be important. Figure 6.2 implies that the majority of parks only have a slight influence upon property prices in the immediate proximity, although this is a reflection of their size. The affect of Bute Park is summarised in Figure 6.3 and has a gentle estimated distance decay curve, as its β-value suggested. Conversely, the affects of proximity to Primary Schools and Sports Centres are much more localised, and have a negligible effect a short distance away.

The geographies of the three main externalities (Bute Park, Heavy Industry and Parks) are illustrated in price surfaces in Figures 6.4 - 6.6 respectively. These were generated in ARC/INFO using the estimated parameters in Model 6.3 to calculate the theoretical impact of each externality upon the properties in the Inner Area. Close inspection of these price surfaces will reveal little pockets of anomalous high and low values. These reflect both the random sample of houses in the Inner Area, and the vagaries of the price surface generating functions ARC/INFO. However, since their impacts are minimal, they can be regarded as white noise.

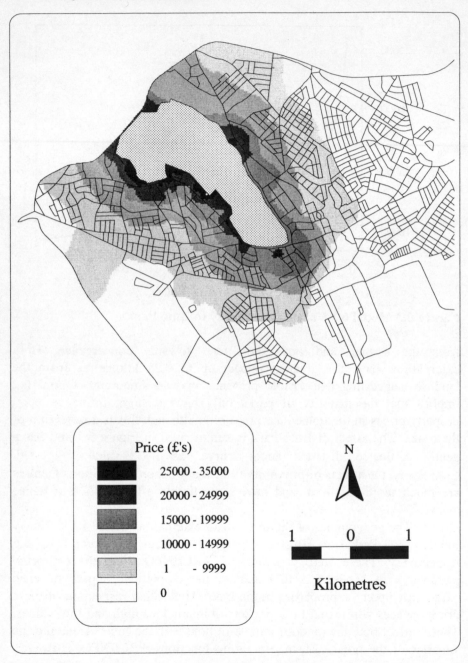

Figure 6.4 Bute Park price surface: model 6.3

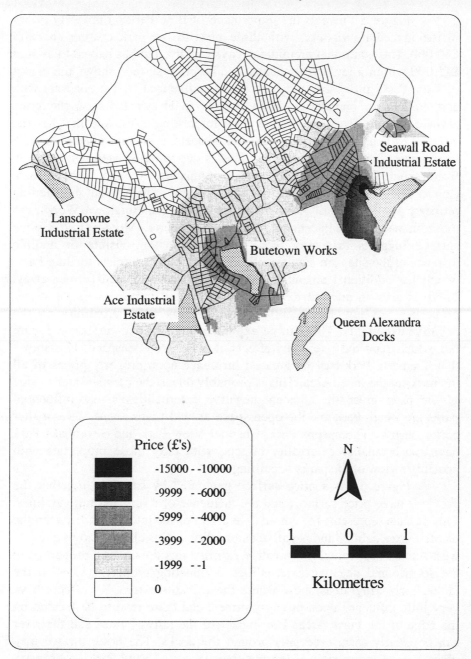

Figure 6.5 Heavy industry price surface: model 6.3

Figure 6.4 reveals the importance of Bute Park on property prices in the immediate vicinity, with those next to the park costing an extra £30,000. This additional premium decays rapidly, and has halved to around £15,000 within a few streets distance. The price surface shows the extent of Bute Parks influence, and how this is influenced by the road network and community boundaries. With respect to Heavy Industry, the price surface is more restricted (Figure 6.5). Interestingly, the site that has the greatest estimated affect in Figure 6.1, Queen Alexander Docks, has the least impact in geographical terms. This is due to its isolated nature, away from immediate residential areas compared to the other sites. Consequently, the sites which has the greatest influence on surrounding property prices, Seawall Road Industrial Estate and Butetown Works, are those immediately adjacent to residential property. Moreover, the price surface suggests that only those properties within visible or audible distance of the sites are significantly affected. This contrasts to Bute Park, which has additional amenity value for properties located further away. There is also an area adjacent to Seawall Road Industrial Estate where properties are not significantly influenced by their proximity to the negative externality. This can be explained by the price surface in Figure 6.6, which maps the externality effects of parks and openspace. This shows that Tremorfa Park has the greatest influence upon property prices of all the parks in the Inner Area. This is probably due to the compensatory value of the park, given the adjacent negative externalities. Other influential parks are Roath Park and the open space around Llansdowne. The smaller parks, such as Thompsons Park, Channel View Park and Sevenoaks Park have much smaller externality effects, with only those properties with possibly a view of the parks benefiting.

Figure 6.7 is a price surface generated by summing together the previous three price surfaces and the influence of rivers and railway lines. This surface represents the cumulative effect of the interaction between the positive externalities and negative externalities. This clearly shows a north-west / south-east split, with positive externalities dominating properties in the former and negative externalities dominating property prices in the latter. Light grey areas show where these major externality effects have very little influence upon property prices, and these tend to be located on the edge of the Inner Area. The impact of the railway lines and the river can be clearly seen, especially around the docks. The price surface map gives a good impression of the positive impact of Bute Park on property prices in the Inner Area. It also demonstrates the complexity of externality effects on a local scale, with positive and negative areas juxtaposed. A

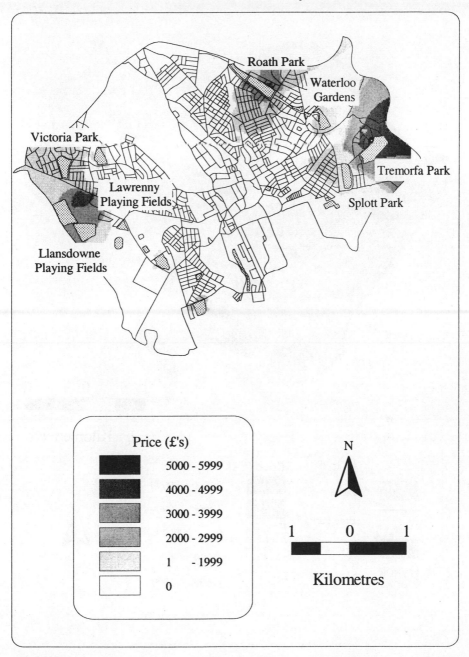

Figure 6.6 Parks and open space price surface: model 6.3

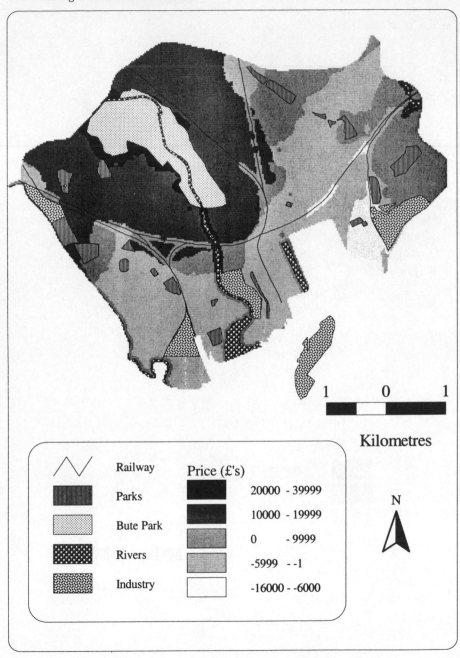

Figure 6.7 Combined externality price surface: model 6.3

Table 6.8 Model 6.4 Street level locational attributes

FIXED

Predictor	Coefficient	Standard Error
Primary Road	-2018	1382.27
Secondary Road	-1463	1153.63
Residential Road	-504	1073.62
Poor 0-50m	-5256	1761.38
Below Ave 0-50m	-5591	1441.17
Above Ave 0-50m	-3227	1357.62
Poor 50-100m	-3724	1382.84
Below Ave 50-100m	-2041	653.30
Above Ave 50-100m	-2582	1911.85
Poor 100-200m	-1553	1063.70
Below Ave 100-200m	-1271	672.49
Above Ave 100-200m	-793	1086.30
Non-Res Buildings	-1521	538.56
Sch: Willows	-6689	3169.08
Sch: Fitzalan	-7064	1800.68
Sch: Cantonia	-4332	2563.56
Sch: Cathays	-3481	3155.04

RANDOM

Parameter	Coefficient	Standard Error
Community Level		
Constant	724	354.86
HCS Area Level		
Constant	262	105.99
Street Level		
Constant	389	156.90
Property Level		
Constant	1503	150.90

-2*(log-likelihood) = -530.331

graphic example of this occurs in Splott, with the distinct split between the properties adjacent to Tremorfa Park, and those next to the industrial estate.

Model 6.4: The Street Level Locational Attributes Model

The affects of street level locational attributes are estimated in Model 6.4 (Table 6.8). These can be summarised as the impact of street quality and the secondary school catchment areas. The street quality effects are interesting. The omitted dummy variables represents 'good' street quality at various distances from the property. Hence, the immediate street quality (up to fifty metres either side of the property - typically an entire street) has a significant impact upon property price and is illustrated in Figure 6.8. This demonstrates that 'poor' and 'below average' street quality has the affect of reducing the property price by around £5,500, compared to 'good' street quality. Interestingly, 'below average' street quality has a more detrimental effect than 'poor' street quality, although this difference is only very marginal. Street quality beyond the immediate property (50-100m and 100-200m) represents the externality effects of the surrounding streets. Model 6.4 indicates that 'below average' and 'poor' street quality within 50-100m of a property are a significant influence upon price, but this is not the case for 'above average' street quality. Street quality beyond 100 metres of the property is generally uninfluential. These results taken together suggest that only when the street quality is below average does it affect price beyond the immediate vicinity of the property (beyond 50m), whilst the distance decay of this effect is steep, and becomes negligible typically within one and half street lengths away from the property. The effect of non-residential buildings in the street also has a significant negative impact upon property, reducing the price by an average of £1,500.

An important result is the influence that living in a specific secondary school catchment area has on property prices. Model 6.4 indicates that, relative to the omitted school variable, only two secondary schools were significantly different and these both had a detrimental effect upon price of around £7,000. However, the size of these estimates casts doubt upon the accuracy of the coefficients. Nevertheless, the model does suggest that school catchment areas may be important in some cases.

The random terms indicate that street level locational attributes reduce street level and HCS Area level variances by over a half. The property level fixed and random terms remain unchanged since the street level locational attributes do not vary at this level. The difference in the

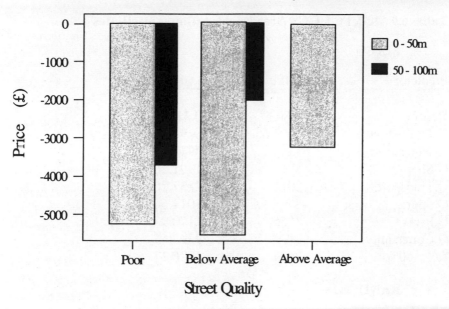

Figure 6.8 Model 6.4 Valuing the relative costs of street quality

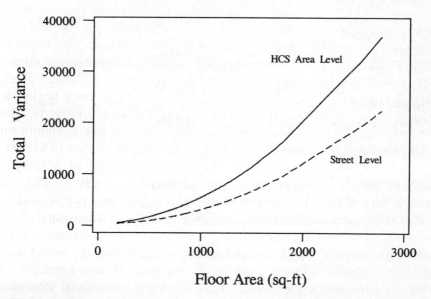

Figure 6.9 HCS Area level and street level variances as a function of floor area

Table 6.9 Model 6.5 HCS Area level locational attributes

FIXED

Predictor	Coefficient	Standard Error
Density	-0.803	0.31
% Opens Space	7073.66	8492.25
% Non-Residential	-114.32	4948.67
Q.Shop	771	1391.26
Q.Transport	2012	2837.76
Q.Sport	-87	1014.66
Q.Parks	-1055	1170.97
Q.Community	713	939.26
LA > 50%	-2684	1926.37

RANDOM

Parameter	Coefficient	Standard Error
Community Level		
Constant	631	310.66
HCS Area Level		
Constant	207	96.072
Street Level		
Constant	389	156.90
Property Level		
Constant	1503	150.90

-2*(log-likelihood) = -540.648

likelihood statistics between this model and Model 6.3 (124) is significant at the 99% level for 10 degrees of freedom, indicating that the street level locational attributes significantly explain the variation in house price.

Model 6.5: The HCS Area Level Locational Attributes Model

Table 6.9 summarises the results of adding HCS Area level locational attributes to Model 6.4. The t-statistics indicate that the only variable that significantly explains the variation in house price is housing density. The remainders are insignificant at the 95% level. This can be explained in part by the lack of detail in the CHCS response data. Moreover, for the majority of the HCS Areas, the types of amenities in question are generally well provided and subsequently their variation will not be of particular importance to the householder. The lack of significance of HCS Areas in which the majority of tenure is Local Authority owned can be attributed to the fact that these represent only 5 of 81 HCS Areas, and thus is of very little significance. Therefore, at the HCS Area Level, the most significant factor influencing property prices are the number of houses per square kilometre. Higher density areas have a negative affect upon property prices compared to lower density areas on the edge of the Inner Area.

The random term suggests only a marginal impact upon HCS Area level and community level variance. However, the difference in the likelihood statistics (10.32) by the addition of this variable is significant at the 99% level with 1 degree of freedom indicating that housing density is a significant factor in explaining property price variation, however small.

Model 6.6: The Community Level Locational Attributes Model

Table 6.10 summarises the addition of community locational attributes. Similar to previous results, social class was significant although the magnitude of the coefficient was only half of that in Model 5.12. This indicates that social class had been acting as a proxy for unaccounted locational attributes. With respect to the random terms, the addition of the social class variable has reduced the variance at the community level by 60%, whilst the difference in the likelihood statistics (14.2) indicate that it significantly explains house price variation at the 99% level with 1 degree of freedom. An examination of the variance at each level suggests that the majority of the unexplained variation now occur at the level of the individual property and the least at the HCS Area level.

Housing Submarkets and Spatial Parameter Drift

Introduction

In the previous models, the housing market was conceived as unified and so the attribute prices remained constant across the Inner Area. Chapter five demonstrated that this was not the case for the whole of Cardiff, with

Table 6.10 Model 6.6 Community level locational attributes

FIXED

Predictor	Coefficient	Standard Error
Social	1415	599.75

RANDOM

Parameter	Coefficient	Standard Error
Community Level		
Constant	241	110.83
HCS Area Level		
Constant	207	96.072
Street Level		
Constant	389	156.90
Property Level		
Constant	1503	150.90

$-2*(\text{log-likelihood}) = -554.892$

submarket conditions causing the implicit price of floor area and social class to vary at higher levels. Hence, it can be hypothesized that the Inner Area housing market will operate under similar conditions. The aim of this section is to ascertain the extent to which submarket conditions influence the valuation of locational externalities. In particular, whether the value of particular externalities are more in some areas than others, and how this relates to the housing stock in these areas. Before this can be achieved, however, the spatial variation in the implicit prices of the structural attributes needs to be modelled.

Spatial Variation in Structural Attributes

Chapter five concluded that the implicit price of floor area varied with community context. To capture this effect, Model 6.6 was re-estimated allowing floor area to vary at the community level. However, the resulting floor area random terms were insignificant. As a result, the model was

Table 6.11 Model 6.7 HCS Area level floor area interactions

RANDOM

Parameter	Coefficient	Standard Error
Community Level		
Constant	203	77.39
HCS Area Level		
Constant	240	100.99
Floor Area	0.0052	0.0016
Floor Area / Constant	0.82	0.30
Street Level		
Constant	408	143.21
Property Level		
Constant	1312	134.89

-2*(log-likelihood) = -580.658

re-estimated allowing floor area to vary at the HCS Area level, producing significant results - see Table 6.11. The random part of Model 6.7 suggests that the unit price of floor area varies between HCS Areas, with the price per square foot being more expensive in HCS Areas with higher than average house prices. This departure from Chapter five, where submarkets were seen to operate at a larger scale, may be explained by the heterogeneous nature of the Inner Area compared to the Cardiff housing market as a whole. Since the housing stock changes across much smaller distances, it may be expected that supply and demand conditions will also vary at this scale. Despite the fact that submarket conditions probably would not operate at such a small scale, floor area was also allowed to vary at the street level. The random part of Model 6.8 (Table 6.12) showed that the co-variance between floor area and average house price did not significantly vary at this level, as was expected, but the unit cost of floor area did. This was confirmed when the model was re-estimated without the covariance term (Model 6.9 - Table 6.13), with the likelihood ratio indicating that the floor area variance term was significant at the 99% level. This unexpected result is discussed later. Figure 6.9 shows how both street and HCS Area variances vary as a function of floor area. The total street level variance is roughly half that of the HCS Area, whilst the relationship between house size and house price variation is much gentler.

Table 6.12 Model 6.8 Street level floor area interactions

RANDOM

Parameter	Coefficient	Standard Error
Community Level		
Constant	186	73.63
HCS Area Level		
Constant	245	99.29
Floor Area	0.0046	0.0016
Floor Area / Constant	0.96	0.31
Street Level		
Constant	343	145.18
Floor Area	0.0027	0.00138
Floor Area / Constant	-0.25	0.26
Property Level		
Constant	1211	129.88

$-2*(\text{log-likelihood}) = -588.011$

Table 6.13 Model 6.9 Re-estimated street level floor area interactions

RANDOM

Predictor	Coefficient	Standard Error
Community Level		
Constant	191	75.78
HCS Area Level		
Constant	259	102.064
Floor Area	0.00449	0.00164
Floor Area / Constant	0.875	0.308
Street Level		
Constant	335	143.66
Floor Area	0.0029	0.00144
Property Level		
Constant	1210	129.97

$-2*(\text{log-likelihood}) = -587.278$

The variation of floor area at the street and HCS Area levels has changed some of the structural attribute estimates. Table 6.14 shows that the implicit price of floor area in detached housing has halved, whilst the

Table 6.14 Model 6.9 The fixed structural attributes

Predictor	Coefficient	Standard Error
Constant	44672	240.26
Floor Area	30.96	1.95
Floor D	4.58	2.22
MT Bath 2	3987	2901.78
Full CH	3346	724.98
Garage	3938	790.56
ORP	2677	877.63
Gdn: None	-5131	1013.94
Gdn: 5-50m	1980	785.26
Needs Mods	-4761	961.11

variable measuring the affect of mid-terraces with two bathrooms has become insignificant. This implies that both these variables had been capturing the spatial variation of the floor area coefficient, compensating for the underestimation of its price in larger properties.

Spatial Variation in Locational Externalities

Introduction The property level locational externalities have previously been shown to have a significant impact upon property prices. Hence, these are also likely to have a differential effect across the housing market. Since the externalities are more likely to vary on a street by street basis, rather than on a HCS Area or community basis, the variance and co-variance terms of the different externalities were added to level two of Model 6.9. The differential impacts of each externality can then be assessed and new prices surfaces generated.

Parks and Open Space Table 6.15 summarises the results of the random part of Model 6.10, which models park and open space externalities. The first thing to note is the insignificance of both the floor area variance term

Table 6.15 Model 6.10 Street level park interactions

RANDOM

Parameter	Coefficient	Standard Error
Community Level		
Constant	163	63.44
HCS Area Level		
Constant	270	95.71
Floor Area	0.0044	0.0013
Floor Area / Constant	0.67	0.26
Street Level		
Constant	389	132.29
Floor Area	0.0006	0.000625
Parks	32798	131192
Parks / Constant	-5.073	0.84
Parks / Floor Area	-0.0093	0.0037
Property Level		
Constant	1347	128.45

$-2*(\text{log-likelihood}) = -587.631$

and the Parks variance term. Their insignificance implies that their implicit prices are constant between streets within an HCS Area. Instead, it is the co-variance terms that are of interest. The likelihood ratio statistic states that the additional two co-variation terms at the street level are significant at the 95% level, when compared to Model 6.7. The negative co-variance between parks and the constant term suggests that marginal impact of parks decreases as average street level property price increases, whilst the covariance term between parks and floor area implies that parks also have a marginally bigger impact upon streets which have smaller housing. Model 6.10 states that parks have greater impact on property prices in streets which have smaller, cheaper property, than in streets which have larger, more expensive property. This is intuitive since smaller, cheaper property tends to be located in higher density areas, and hence access to open space will be more valued than in areas where the housing density is lower. Furthermore, this result implies that it is not the implicit price of parks that is important, rather it is the impact it has on the structural attributes. The price per unit of floor area would appear to drift with respect to proximity to parks and open space. This is illustrated in Figure 6.10, which shows that the price surface for parks and open space has altered such that it has become relatively more expensive in the higher density housing around Victoria Park, Lansdowne Park and Roath Park, and relatively less expensive around Tremorfa Park. The overall magnitude of the effect has also declined by a third, indicating that the previous estimate was compensating for the differential effects of house size.

Heavy industrial sites Table 6.16 summarises the random part of Model 6.11, the street level variation of proximity to heavy industrial sites. In comparison to Model 6.10, all the heavy industry random terms are insignificant, whilst the floor area variance term remains unchanged from Model 6.9. The likelihood ratio states that the additional random terms have had no significant effect on explaining street level variation. Hence, it can be concluded that proximity to heavy industrial sites has a constant effect across the Inner Area, regardless of house price and property size.

Table 6.16 Model 6.11 Street level heavy industry interactions

RANDOM

Parameter	Coefficient	Standard Error
Community Level		
Constant	191	74.61
HCS Area Level		
Constant	269	103.94
Floor Area	0.0044	0.0016
Floor Area / Constant	0.87	0.307
Street Level		
Constant	330	143.12
Floor Area	0.0028	0.001411
Heavy Ind	8557	18307.77
Heavy Ind / Constant	1.67	3.37
Heavy Ind / Floor Area	0.00640	0.0174
Property Level		
Constant	1216	130.24

-2*(log-likelihood) = -588.255

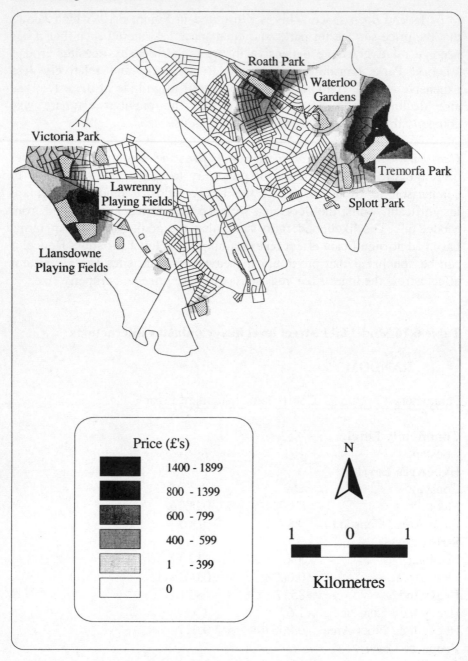

Figure 6.10 Parks and open space price surface: model 6.10

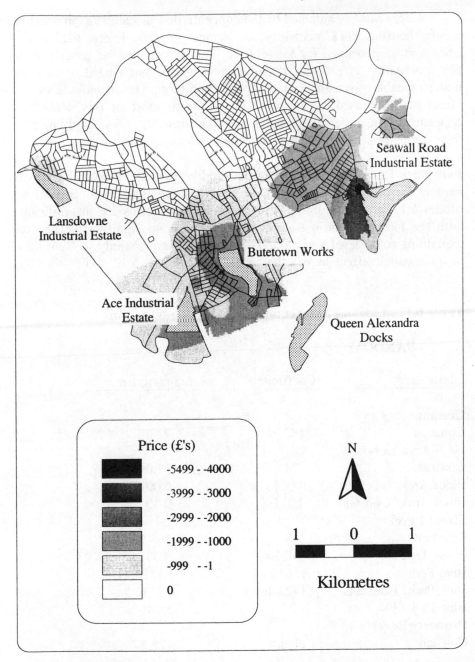

Figure 6.11 Heavy industry price surface: model 6.11

This can be explained by the concentration of such sites in areas of similar housing stock, namely small terrace property. Figure 6.11 is the price surface generated for Model 6.11. This shows that the geography of the externality effect has remained essentially unchanged, which is understandable given the lack of spatial variation. The magnitude of the effect has decreased by around 60%, although most of this decline is concentrated in the areas immediately adjacent to the sites, implying that the externality effect is much gentler than previously estimated.

Bute Park Table 6.17 summarises the random part of Model 6.12, which models the street level variance of proximity to Bute Park. Similar to Model 6.11, all the additional Bute Park random terms are insignificant, with the likelihood ratio stating that they have no significant effect on explaining street level variation. This implies that proximity to Bute Park has a constant effect across the Inner Area. Figure 6.12 shows that the

Table 6.17 Model 6.12 Street level Bute Park interactions

RANDOM

Parameter	Coefficient	Standard Error
Community Level		
Constant	182	72.38
HCS Area Level		
Constant	221	95.09
Floor Area	0.0046	0.0017
Floor Area / Constant	0.911	0.30
Street Level		
Constant	348	168.26
Floor Area	0.0030	0.00147
Bute Park	15613.65	34710.91
Bute Park / Constant	1423.40	1723.35
Bute Park / Floor Area	-1.75	5.14
Property Level		
Constant	1191	128.27

-2*(log-likelihood) = - 590.854

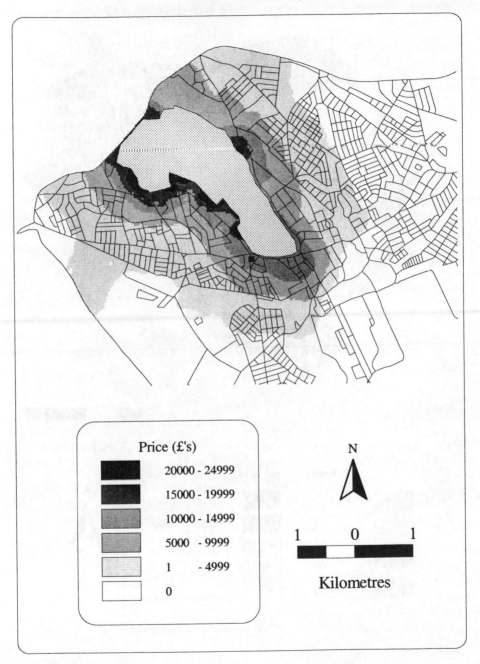

Figure 6.12 Bute Park price surface: model 6.12

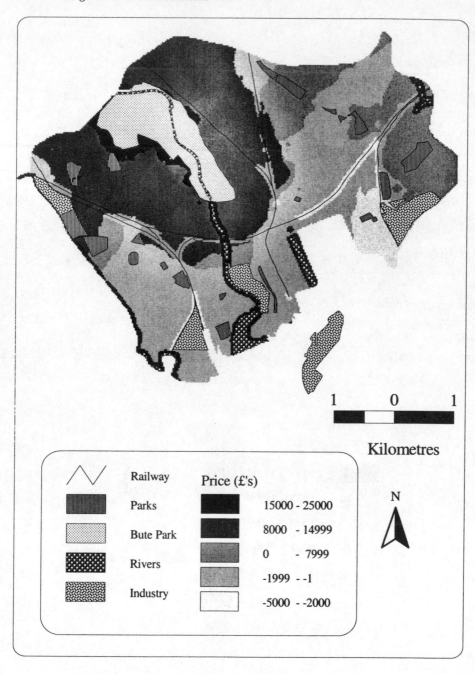

Figure 6.13 Combined spatial externality price surface

influence of Bute Park on the surrounding property prices has decreased by 25%, although again, this decline is greatest in areas immediately adjacent to the Park.

Combined spatial externality effects Figure 6.13 is a summary of the combined externality effects of Model 6.12. It can be seen that the main consequence of allowing attribute prices to vary across the Inner Area has been to reduce the impact of the externalities at the extremes and to make their influence subtler. Therefore, the price surface no longer represents a simple mirror image of the externality, with prices simply decreasing with distance. Instead, the spatial variation has created a mosaic of prices, reflecting both the strength of the externality and the size of the property. For instance, the influence of Bute Park has declined, such that only the large properties located in Llandaff and Riverside are significantly affected. The areas where overall externality effects are minimal have also increased, and these help highlight localised externality effects such as Roath Park and Victoria Park. The impact of railway lines has also become negligible, with the greatest influence upon property prices occurring in Adamsdown and Splott.

Implicit prices of locational attributes Table 6.18 summarises the estimated implicit prices of the locational attributes for the above price surface. This shows that the greatest changes have occurred with respect to property level externalities. As has already been demonstrated, the implicit prices of parks and open space have almost halved, whilst the implicit price of the heavy industrial area externality has decreased by around 60% to £9,400. The influence of Bute Park has also decreased by 25%. Only the externality effects generated by primary schools and sports facilities have remained unaffected, although this is due to the very short distance over which they operate. Street quality has become more important, particularly between fifty to one hundred metres from the property, where 'above average' street quality has become significant. The significance of the school catchment areas has declined, although they still represent an important determinant of property price. At the HCS Area level, the importance of housing density has declined now that the differential price of floor area has been accounted for, whilst the affect of social class at the community level remains substantially unchanged.

Table 6.18 Model 6.12 Locational attribute parameter estimates

Parameter	Coefficient	Standard Error
Dist MWAY	-2.23	0.89
Bute Park(β = 0.25)	8915	14530.11
Parks (β = 0.25)	82573	29385.0
Heavy Ind(β = 0.5)	-9472	4065.23
Sports(β = 2.0)	0.0002573	0.00013
Primary(β = 3)	0.000045	0.000022
Rail 0-50m	-1045	406.60
River 0-50m	10704	2504.59
Poor 0-50m	-6433	1613.51
Below Ave 0-50m	-6927	1327.89
Above Ave 0-50m	-4881	1252.04
Poor 50-100m	-5783	1593.11
Below Ave 50-100m	-5598	1523.15
Above Ave50-100m	-4773	1485.18
Non-Res Buildings	-1708	527.06
SCH: Willows	-4721	2097.78
SCH: Fitzalan	-5968	1334.23
Sch: Cantonia	-3046	1821.00
Sch: Cathays	-1179	2264.82
Density	-0.625	0.29
Social	1317	520.81

Conclusions

The aim of this chapter was to show how locational attributes influence house prices, and how these vary within the built environment. By building a multi-level hedonic model for the Inner Area of Cardiff, the locational attributes were allowed to vary at different spatial levels. This permitted the locational attributes to enter the house price determination process at the correct level. It also allowed some locational attributes to impact upon property prices to a greater extent than others did. Generally, it can be concluded that locational attributes decline in their importance the higher the level that they operate, with property level attributes having the greatest influence upon property prices. Buyers and sellers of housing are more concerned with the immediate compositional attributes of the property, as opposed to the more contextual locational attributes of the community.

Different externalities have been shown to have different impacts on the built environment, and that positive externalities, such as parks, have a wider influence than negative externalities once the size of the effect has been taken into consideration. It was also demonstrated that some externalities, such as primary schools, operate only over very small distances, which implies that proximity in terms of walking distance may be important. With respect to street quality, it is not only the immediate street quality that impacts upon property price, but also neighbouring street quality. However, the distance decay associated with neighbouring street quality would appear to be greater.

An important outcome of this chapter has been the exploration of how the impact of locational externalities varies across the Inner Area, and how they interact with the housing stock. It can be concluded that the implicit price of locational externalities do not vary *per se*, rather the model would appear to indicate that they influence the implicit prices of the structural attributes, and the cost of floor area in particular. This results in a complex interaction of structural and locational, with areas of positive and negative externality effects in juxtaposition. This is especially interesting since it illustrates how externality effects operate over very localised areas. It can therefore be concluded that the influence of locational attributes upon property prices is complex and less obvious than structural attributes.

7   Conclusion

Introduction

Valuing the built environment is a non-trivial exercise. It involves an understanding of both housing market dynamics and the effects of locational externalities upon house prices. This book has demonstrated techniques that enable both tasks to be accomplished within a GIS framework using spatially sensitive specifications of the hedonic house price function. The purpose of this conclusion is to draw together some of the key issues of the research and evaluate the empirical results with respect to the micro-economic theories of residential location and housing markets. Integral to this is an appreciation of the role of GIS in the integration and visualization of the data, and crucially, in the spatial analysis of the locational attribute data.

Valuing the Built Environment

Micro-Economic Theory of Housing Markets

The 1960s and the 1970s saw a proliferation of micro-economic theories and mathematical models of housing markets, residential location and landuse in both the UK and USA (e.g. Muth, 1969, Batty, 1976). These were essentially demand led models that assumed that western capitalist housing markets operated under Pareto Optimum conditions. In such a formulation, the supply and demand of housing was assumed to be in perfect equilibrium. Furthermore, the supply of housing was either ignored or was assumed to be effortless adapted to variations in demand. Deductions from these models were used to explain the spatial patterning of residential location, which centred on the concept of a bid-rent function. This function described the process in which households' traded-off accessibility to the city centre with housing attributes, and in particular housing size. The result of this trade-off between accessibility and space was a decrease in rents from the city centre outwards, which declined at a decreasing rate.

However, since these models were essentially descriptive, their deductions required verification from empirical evidence. The motivation behind the early hedonic house price research was to provide this empirical evidence, namely in the estimation of the negative rent gradient which would strengthen the argument for the accessibility / housing size trade-off. However, the estimation of the negative rent gradient has been a cause of controversy in hedonic house price studies, with counter-intuitive or insignificant results being the norm. In addition, hedonic house price research has questioned the validity of the concept of a market in perfect equilibrium functioning under Pareto Optimum conditions. Instead, hedonic research has become increasingly concerned with disequilibrium, that is the supply and demand of housing operating under restrictive conditions. Initially, the main emphasis of this research was upon racial segregation in North American cities, and how this affected the housing market. In recent years though, housing market stratification in more general terms has became a predominant concern. However, similar to the negative rent gradient hypothesis, the evidence for the existence of submarkets has also been contradictory. Therefore, both this interest in housing market disequilibrium and the lack of consistent evidence for a negative rent gradient contributed to micro-economic housing market theories falling out of favour in the late 1970s.

The research in Cardiff has demonstrated that this may not be the case. There is little doubt that previous hedonic price functions have been misspecified, whilst many studies have used poorly specified data, particularly with respect to locational attributes. Together, these factors may have been responsible for the lack of consistent empirical evidence of the neo-classical urban models. By using a highly disaggregated database, geo-referenced to a higher resolution than many previous studies, a negative rent gradient has been identified for Cardiff. Moreover, this rent gradient becomes increasingly concave as the hedonic specification became more sensitive to the spatial dynamics of the housing market. This would seem to verify, with respect to Cardiff, the concept of a trade-off between access to the city centre and all of the other housing attributes.

The research also demonstrated the existence of submarkets, implying housing market disequilibrium. Separate submarkets for detached housing can be explained in terms of supply and demand. In terms of supply detached housing has, on average, the largest amount of living space of all the dwelling types in Cardiff. It also tends to be restricted to suburban locations, particularly in areas that have good neighbourhood quality. In terms of demand, detached housing would be desired by larger households. In addition, high neighbourhood quality would also make detached housing particularly appealing. This interaction between supply and demand would appear to have created specific market conditions, resulting in a separate submarket for detached housing.

Submarkets operating within defined geographical areas, in this case communities, can be explained by factors such as imperfect knowledge of the housing market, a desire to live near family and friends, or restrictions placed upon the supply of housing, such that perfect substitution is not possible. In addition, institutions and actors operating within the housing market, such as estate agents, may also reinforce these submarket conditions. Estate agents use the twenty-six Cardiff communities as a basis for structuring house sales. For instance, houses for sale were grouped into their respective communities, which help to guide potential buyers. More importantly, since estate agents value property using the comparative method (Millington, 1990), communities would become integral to the valuation process. Therefore, it is hardly surprising that property prices reflect community boundaries. This is shown in spatial autocorrelation maps that indicate that house price residuals within a community are more similar than house price residuals in different communities.

Hence the results of this research have several implications for the micro-economic theory of housing markets and residential location. The main implication is the estimation of the negative rent gradient that gives support to the concept of the bid-rent function and the trade-off between accessibility to the city centre and house size. However, the evidence of housing submarkets would appear to contradict the assumption of a housing market equilibrium that is fundamental to the micro-economic theory. In particular, the research points to the possible importance of estate agents as structuring the valuation process and hence to some extent influencing the spatial dynamics of the housing market.

Hedonic House Price Research

The hedonic pricing method is very well established, particularly in North America. From its origins as a method of producing empirical evidence to underpin the micro-economic theory of housing markets, it has subsequently been used as a common method of imputing the value of housing attributes. It has especially been employed to impute environmental benefits and are regularly used in demand equations to estimate the costs of specific externalities, such as the costs of air and noise pollution. Substantial intellectual energy has been expended on the specification of these demand equations, with comparably very little on the specification of the hedonic price function. It is now clear that the traditional specification is far from robust with respect to the incorporation of space. The corollary of this is that the estimates of the hedonic price function may be biased and inefficient, and that this inefficiency will subsequently enter the demand equations. If the results of these demand equations are used to inform policy making then the consequences of this misspecification becomes non-trivial.

An investigation into the ability of three hedonic specifications to model the spatial structures of the housing market have indicated that the traditional specification produces the most inefficient models with respect to heteroscedasticity and spatial autocorrelation. With respect to these criteria, the best specification was the multi-level specification. Conceptually, this specification also best describes housing market dynamics once geographical factors are taken into consideration. This specification allows the spatial structures inherent in the valuation of property, such as community boundaries, to be modelled explicitly, whilst the specification also allows both the compositional effects of the housing stock and the contextual effects of location to be modelled simultaneously. However, it can be argued that the multi-level specification places a too rigid a criterion upon the delimitation of spatial boundaries within the housing market, and does not allow for possible spill over effects between them. In this respect, the spatial parameter drift specification is conceptually more appealing, although as was shown, the inability of this specification to model the spatial variation in locational attributes caused heteroscedasticity in the estimated parameters. Therefore, it can be concluded that the multi-level hedonic specification is the most efficient at modelling housing market, and this should have implications for future research.

Locational Externalities Effects

Locational externality effects have been a common concern in the urban economic literature and a particularly important concept of how the built environment is valued. It has traditionally been couched in terms of power and conflict, specifically in how negative externalities impact upon peoples' lives and property values, and how people subsequently come together to exclude them from their locality. Although these incidents are perhaps overstated, it is sill the case that in recent years, NIMBY (Not In My Back Yard) -ism has remained an important issue within the UK. This research, and particularly the analysis in chapter six, has demonstrated that locational attributes are a very important part of the house price valuation process. Moreover, is has been shown that higher income households may value locational attributes marginally more than structural attributes, and hence that these may represent a larger proportion of the value of higher priced properties. This was supported by the findings that the values of structural attributes are intimately bound up with locational attributes. If this is the case then it gives credence to the argument that higher income households are more likely to come together in action to protect their property prices, since locational externalities represent a larger investment in the price of their property.

The complex geography of locational externality effects at the local level was also demonstrated. Positive and negative externalities are juxtaposed across very small scales, and their influence upon property prices can be measured on a street-by-street basis. This has important implications for hedonic house price research, since much previous work has been undertaken at much lower resolutions. More specifically, any estimation of the value of amenities will have to take these small-scale variations into account. This has joint implications with the previous discussions concerning the specification of the hedonic house price function. Both the ability to model spatial effects and the resolution of the data are important if efficient estimates of housing attributes are to be made.

GIS and House Price Analysis

Any questions concerning the resolution of house price data and the measurement of locational attributes has implications with the role of the GIS within hedonic research. In the past, the role of GIS in spatial analysis has been questionable. Commentators such as Openshaw (1995) have argued that GIS should move away from Geographic Information Handling to Geographic Information Using, and that GIS has been under utilised as a powerful tool in spatial analysis, especially in the social sciences. One of the problems contributing to this under utilisation has been a lack of the availability of spatially disaggregated socio-economic data, geo-referenced to a high resolution. However, with the advent of digital products such as ADDRESS-POINT, and appropriate matching techniques, the problems of geo-referencing socio-economic data to a high resolution are being slowly addressed. Large and complex addressed based datasets, such as the CHCS, are now capable of being linked and manipulated at a level of disaggregation not possible before. Indeed, this research has constructed and utilised a GIS at a level of disaggregation and complexity not used before in any hedonic house price study, with the housing data having been modelled at the appropriate level of resolution for the first time. It can therefore be concluded that it is the role of the GIS in the research, and the high resolution of the data, which are in no doubt responsible for the quality of the results.

The research demonstrated the continued importance of postcodes as the basis of matching address-based datasets. Such datasets are becoming increasingly important with the advance of geo-marketing and private companies collecting information upon individual people. By highlighting the procedural problems with address-based matching it can be concluded that the standardisation of addresses should be paramount, and that this is now increasingly possible with advent of the British Standard BS7666.

The importance of the GIS as a visualization medium was proved by the locational externality price surfaces. Whereas line graphs illustrated the magnitude of the effects, it was the GIS that depicted their complex geography. The price surfaces demonstrated that in heterogeneous urban areas, externality effects vary in scale and magnitude across very small areas, creating an intricate geography of house price variation. Using visualization to depict complex socio-economic phenomena is a neglected area in the social sciences, despite the fact that visualization tools and technologies have been generally available for some time (Orford, et al. 1999).

The Future

This book has set out to investigate how the built environment is valued at the local level from an inherently geographical (as opposed to a wholly economic) perspective. By doing so, it has highlighted some of the potential problems faced when modelling large and complex spatial datasets. Furthermore, the research has demonstrated the importance of GIS in structuring and manipulating the data. Indeed, it is the ability of the GIS to handle large and complex socio-economic datasets at a high level of disaggregation, and to generate locationally sensitive externality data using the spatial analysis tools, that allowed this research to produce such detailed results at such a small scale. For instance, since GIS is a perfect medium for handling housing attribute and valuation data, it has increasingly been used by real estate agents to aid their business, especially in North America (Dixon, 1992). There is therefore a good argument for its introduction into the UK. Property valuation is at best a very inexact science (Millington, 1990), due in part to a lack of available, comparable comprehensive databases. The introduction of information technology, and GIS in particular, may go some way to alleviate the present uncertainties in the valuation procedure.

Future research may be able to build on this case study, improving the GIS and spatial analysis techniques to unravel the complexities of the built environment at even finer levels of resolution. Together with the availability of increasingly sophisticated datasets, the new specifications of the hedonic house price function and the related spatial econometric applications have the potential of opening up new avenues of research into the built environment that has just not been possible before.

Bibliography

Adair, A.S., Berry, J.N. and McGreal, W.S. (1996), 'Hedonic modelling, housing submarkets and residential valuation', *Journal of Property Research.* Vol 13. pp. 67-83.

Alonso, W. (1964), *Location and Land Use,* Harvard University Press, Cambridge, MA.

Anas, A. and Dendrinos, S.D. (1976), 'The new urban economics: A brief summary', in George J. Papageorgiou (ed.), *Mathematical Land Use Theory,* Lexington Books.

Anscombe, F.J. (1973), 'Graphs in statistical analysis', *The American Statistician,* Vol. 27. pp. 17-22.

Anselin, L. (1988a), *'Spatial Econometrics: Methods and Models',* Dordrecht: Kluwer Academic.

Anselin, L (1988b), 'Lagrange multiplier test diagnostics for spatial dependence and spatial heterogeneity', *Geographical Analysis.* Vol. 20.1. pp. 1-17.

Anselin, L. (1992), 'Spatial dependence and spatial heterogeneity, model specification issues in the spatial expansion paradigm', in E. Cassetti and J.P. Jones (eds), *Applications of the Expansion Method.* London, Routledge.

Anselin, L and Griffith, D.A. (1988), 'Do spatial effects really matter in regression analysis?' *Papers of Regional Science Association,* Vol 65. pp. 11-34.

Bailey, T.C. (1994), 'A review of statistical spatial analysis in Geographical Information Systems', in S. Fotheringham and P. Rogerson (eds), *Spatial Analysis and GIS,* Taylor and Francis Ltd. pp. 13-44.

Bajic, V. (1984), 'An analysis of the demand for housing attributes', *Applied Economics,* Vol 16. pp. 597-610.

Ball, M. (1973), 'Recent empirical work on the determinants of relative house prices', *Urban Studies, Vol. 10.* pp. 213-233.

Ball, M (1974), 'The determinants of relative house prices: A reply', *Urban Studies.* Vol 11. pp. 231-233.

Ball, M. (1985), 'The urban rent question', *Environment and Planning A,* Vol. 17.4. pp. 503-525.

Ball, M. and Kirwan, R. (1977), 'Accessibility and constraints in the urban housing market', *Urban Studies.* Vol. 14. pp. 11-32.

Bassett, K. and Short, J. (1980), *Housing and Residential Structure:*

Alternative Approaches, Routledge and Kegan Paul: London.

Batty, M. (1976), *'Urban Modelling: Algorithms, Calibrations, Predictions',* Cambridge University Press.

Batty, M. (1994). 'Urban models twenty five years on', *Environment and Planning B.* Vol. 21.5. pp. 515-516.

Batty, M. and Longley, P. (1987), 'A fractal based description of urban form', *Environment and Planning B.* Vol. 14.2. pp. 123-134.

Batty, M. and Xie. Y.(1994a), 'Modelling inside a GIS 1: Model structures, exploratory data analysis and aggregation', *International Journal of GIS.* Vol. 8.3. pp. 291-307.

Batty, M. and Xie. Y. (1994b), 'Modelling inside a GIS 2: Selecting and calibrating urban models using Arc-Info', *International Journal of GIS.* Vol. 8.5. pp. 451-470.

Beaverstock, J., Leyshon, A., Rutherford, T., Thrift, N., and Williams, P. (1992), 'Moving houses: the geographical reorganisation of the estate agency industry in England and Wales in the 1980s.', *Transactions, Institute of British Geographers.* Vol 17. pp. 166-182.

Beckman, M.J. (1968), *Location Theory,* Random House, Inc.

Belsley, D.A., and Kuh, E. and Welsch, R.E. (1980), 'Robust estimation of an hedonic housing price equation', *Regression Diagnostics.* Chapter 4.4. pp 229-245.

Bender, B. and Hwang, H. (1985), 'Hedonic house price indices and secondary employment centres', *Journal of Urban Economics.* Vol. 17. pp. 90-107.

Berry, B.J. (1976), 'Ghetto expansion and single family housing prices: Chicago 1968-1972', *Journal of Urban Economics.* Vol. 3. pp. 397-423.

Blake, M and Openshaw, S. (1996), School of Geography, Leeds University, Leeds. LS2 9JT *http://www.geog.leeds.ac.uk/staff/m.blake/v-sel/v-sel.htm*

Boddy, M. (1980), *'The Building Societies ',* Macmillan, London.

Bourne, L.S. (1981), *'The Geography of Housing.',* London: Edward Arnold.

Boyle, P. and Dunn, C.E. (1991), 'Redefinition of Enumeration District centroids: a test of their accuracy by using Thiessen Polygons', *Environment and Planning A. Vol. 23.* pp. 1111-1119.

Brown, G.M. and Pollakowski, H.O. (1977), 'The economic valuation of shore line', *Review of Economics and Statistics.* Vol. 59. pp. 272-278.

Butler, R. (1982), 'The specification of hedonic indexes for urban housing', *Land Economics.* Vol. 58. pp. 98-108.

Can, A. (1990), 'The measurement of neighbourhood dynamics in urban house prices', *Economic Geography.* Vol. 66. pp. 254-272.

Can, A. (1992), 'Specification and estimation of hedonic house price models', *Regional Science and Urban Economics*. Vol. 22. pp. 453-474.

Cassetti, E. (1972), 'Generating models by the expansion method: Applications to geographical research', *Geographical Analysis*. Vol. 4. pp. 81-91.

Cassetti, E. (1992), 'The dual expansion method: An application for evaluating the effect of population growth on development', in E.Cassetti and J.P.Jones (eds), *Applications of the Expansion Method*, London, Routledge.

Chambers, J.M., Kleiner, B. and Tukey, P.A. (1983), *'Graphical Methods For Data Analysis'*, Duxbury Press, Boston. pp. 258-259.

Chatterjee, S. and Price, B. (1977), *Regression Analysis by Example*, John Wiley and Sons.

Cheshire, P. and Sheppard, S. (1995), 'On the price of land and the value of amenities', *Econometrica*. Vol. 62.246. pp. 247-267.

Cliff, A.D. and Ord, J.K. (1981), *Spatial Processes: Models and Applications*, Pion, London.

Collins, A. and Evans A. (1994), 'Aircraft noise and residential property values: An artificial neural network approach', *Journal of Transport Economics and Policy*, Vol. 27-28. pp. 175-198.

Cooley, R., Hobbs, M., and Clewer, A. (1995), 'Location and residential property values', in M. Fischer, T. Sikos and L. Bassa (eds), *Recent Developments in Spatial Information, Modelling and Processing* Geomarket Co.

Cox, K.R. (1979), *Location and Public Problems*, Basil Blackwell: Oxford.

Cropper, M.L., Deck, L.B., McConnell, K.E. (1988), 'On choice of functional form for hedonic price functions', *Review of Economics and Statistics*. Vol. 70.4. pp. 668-675.

Cubin, J.S. (1970), 'A hedonic approach to some aspects of the Coventry housing market', *Warwick Economic Research Papers* 14.

Cushnie, J. (1994), 'A British standard is published', *Mapping Awareness*. Vol. 8.5. pp. 40-43.

Daniels, C.B. (1975), 'The influence of racial segregation on housing prices', *Journal of Urban Economics*. Vol 2. pp. 105-122.

Darling, A.H. (1973), 'Measuring benefits generated by urban water parks', *Land Economies* Vol. 49. pp. 22-34.

Daultrey, S. (1976), *Principal Components Analysis*, Concepts and Techniques in Modern Geography 8 Norwich: Geo Books.

Daunton, M.J. (1977), *Coal Metropolis: Cardiff 1870-1914*.

Davies, G. (1974), 'An econometric analysis of residential amenity', *Urban Studies*. Vol. 11. pp. 217-226.

Dear, M.J. (1976), 'Spatial externalities and locational conflict' in D. Massey and P. Batey (eds), *London Papers in Regional Science: Alternative Framework for Analysis,* London, Pion. pp. 152-167.

DeLisle, J.R. (1984), 'Market segmentation: implications for residential appraisal', *The Real Estate Appraiser and Analyst.* Vol. 50. pp. 48-54.

Department of the Environment (1991), *The English Housing Condition Survey,* London: HMSO.

Dinan, T.M. and Miranowski, J.A. (1989), 'Estimating the implicit price of energy efficiency. improvements in the residential housing market: A hedonic approach', *Journal of Urban Economics.* Vol 25. pp. 52-67.

Dixon, T. (1992), 'I.T. Applications in property: An overview and future trends', *Mapping Awareness* Vol 6.2.

Do, A.Q. and Grunditski, G. (1995), 'Golf-course and residential house prices - an empirical examination', *Journal of Real Estate Finance and Economics.* Vol. 10.3. pp. 261-270.

Do, A.Q., Wilbur, R. and Short, J. (1994), 'An empirical examination of the externalities of neighbourhood churches on housing values', *Journal of Real Estate Finance and Economics.* Vol. 9. pp. 127-136.

Dorling, D. (1995), *A New Social Atlas of Britain,* John Wiley and Sons.

Duban, R.A. and Sung, C. (1987), 'Spatial variation in the price of housing. Rent gradients in non-monocentric cities', *Urban Studies.* Vol 24. pp 193-204.

Duban, R.A. and Sung, C. (1990), 'Specification of hedonic regressions: non-nested tests on measures of neighbourhood quality', *Journal of Urban Economics.* Vol 27. pp. 97-110.

Dunn, R. (1989), 'Building regression models: the importance of graphics', *Journal of Geography in Higher Education.* Vol. 13.1, pp. 15-30.

Dunn, R., Longley, P.A., and Wrigley, N. (1987), 'Graphical procedures for identifying functional form in binary discrete choice models - a case study of revealed tenure choice', *Regional Science and Urban Economics,* Vol. 17.1. pp. 151-171.

Ellickson, B. (1981), 'An alternative test of the hedonic theory of housing markets', *Journal of Urban Economics,* Vol 9. pp. 56-79.

Evans, A.W. (1991), 'Rabbit hutches on postage stamps - planning development and political-economy', *Urban Studies.* Vol. 28.6. pp.853-870.

Evans, A.W. (1973), *The Economics of Residential Location,* London: Macmillan.

Fleming, M.C. and Nellis, J.G. (1981), 'The interpretation of house price statistics for the UK', *Environment and Planning A*. Vol 13. pp 1109-1124.

Fleming, M.C. and Nellis, J.G. (1985), 'Research policy and review 2. House price statistics for the United Kingdom: A survey and critical review of recent developments', *Environment and Planning A*. Vol 17. pp 297-318.

Fleming, M.C. and Nellis, J.G. (1987), *Spon's House Price Data Book*, London: Spon.

Fleming, M.C. and Nellis, J.G. (1992), 'Development of standardised indices for measuring house price inflation incorporating physical and locational characteristics', *Applied Economics*. Vol 24. pp. 1067-1085.

Fleming, M.C. and Nellis, J.G. (1994), 'The measurement of UK house prices: A review and appraisal of the principal sources', *Housing Finance*. No. 24.

Flowerdew, R. (1991), 'Spatial data integration', in D.J. Maguire, M.F. Goodchild and D.W. Rhind (eds), *Geographical Information Systems: Methodology and Potential Applications,* Longmans, London.

Flowerdew, R. and Openshaw, S. (1987), 'A review of the problems of transferring data from one set of areal units to another incomplete set', *Research Paper 4. Northern Regional Research Laboratory*, Lancaster and Newcastle.

Follain, J.R. and Jimenez, E. (1985), 'Estimating the demand for housing characteristics: A critique and survey', *Regional Science and Urban Economics*. Vol 15. pp. 77-107.

Forrest, D., Glen, J. and Ward, R (1996), 'The impact of a light rail system on the structure of house prices - A hedonic longitudinal study', *Journal of Transport Economics and Policy*. Vol. 30.1. pp. 15-37.

Fortney, J (1996), 'A cost-benefit locational-allocation model for public facilities - An econometric approach', *Geographical Analysis*. Vol. 28.1 pp. 67-92.

Foster, C.D., Jackman, R.A. and Perlman, M. (1980), '*Local Government Finance in a Unitary State',* George Allen and Unwin: London.

Foster, S.A. (1991), 'The expansion method: Implications for geographical research', *Professional Geographer*. Vol. 43.2. pp. 131-142.

Fotheringham, A.S. and Wong, D.W.S (1991), 'The Modifiable Areal Unit Problem in multivariate statistical analysis', *Environment and Planning A*. Vol. 23.7 pp. 1025-1044.

Fox, J. (1991), *Regression Diagnostics*, Sage Publications.

Freeman, A.M. (1979a), *'The Benefit of Environmental Improvement: Theory and Practice'* John Hopkins University Press. Baltimore and London.

Freeman, A.M. (1979b), 'Hedonic prices, property values and measuring environmental benefits: A survey of the issues', *Scandinavian Journal of Economics.* Vol. 81. pp. 154-173.

Freeman, R. and Dixon, T. (1992), *'Information Technology Applications in Commercial Property',* MacMillian Press Ltd.

Garrod, G. and Willis, K. (1992a), 'Valuing the goods characteristics - an application of the hedonic price method to environmental attributes', *Journal of Environmental Management.* Vol. 34.1 pp. 59-76.

Garrod, G. and Willis, K. (1992b), 'The environmental economic impact of woodland: A two stage hedonic price model of the amenity value of forestry in Britain', *Applied Economics.* Vol. 24. pp. 715-728.

Gatrell, A.C. (1989), 'On the spatial representation and accuracy of address-based data in the United Kingdom', *International Journal of Geographical Information Systems.* Vol 3. pp. 335-348.

Gatrell, A.C., Dunn, C.E., and Boyle, P.J. (1991), 'The relative utility of the Central Postcode Directory and PinPoint Address Code in applications of Geographic Information Systems', *Environment and Planning A.* Vol. 23. pp. 1447-1458.

Getis, A. (1994), 'Spatial dependence and heterogeneity and proximal databases', in Fotheringham, S. and Rogerson, P. (eds), *Spatial Analysis and GIS,* Taylor and Francis Ltd. pp. 105-120.

Gillard, Q. (1981), 'The effect of environmental amenities on house values: the example of a view lot', *Professional Geographer* Vol. 33.2. pp. 216-220.

Goldstein, H. (1987), *Multi-level Models in Educational and Social Research,* Charles Griffin, London.

Goldstein, H. (1995), *Multi-level Statistical Models,* Kendals Library of Statistics 3, Second Edition.

Goodchild, M.F. (1986), *Spatial Autocorrelation,* Concepts and Techniques in Modern Geography 47, Norwich: Geo Books.

Goodchild, M.F. (1987), 'A spatial analytical perspective on Geographical Information Systems', *International Journal of Geographical Information Systems.* Vol. 1. pp. 335-354.

Goodchild, M.F (1988), 'Towards an enumeration and classification of GIS', in R.T. Aangeenburg and Y.M. Schiffman (eds), *International Geographic Information Systems (IGIS) Symposium: The Research Agenda.* pp. 67-77.

Goodchild, M.F. (1991), 'The technological setting of GIS' in D.J. Maguire, M.F. Goodchild and D.W.Rhind (eds), *Geographical Information Systems: Methodology and Potential Applications,* Longmans, London. pp. 45-54.

Goodman, A.C. (1978), 'Hedonic prices, price indices and housing markets', *Journal of Urban Economics.* Vol. 5. pp. 471-484.

Goodman, A.C. (1979), 'Externalities and non-monocentric price distance functions', *Land Economics,* Vol 16. pp. 321-328.

Goodman, A.C. (1981), 'Housing submarkets within urban areas - definitions and evidence', *Journal of Regional Science.* Vol. 21.2. pp. 175-185.

Goodman, M. (1990), 'Players are hungry for a slice of cake', *Money Weekly* 7. Feb. pp. 9.

Gordon, P., Richardson, H.W. and Wong, H.L. (1986), 'The distribution of population and employment in a polycentric city: the case of Los Angeles', *Environment and Planning A.* Vol. 18. pp. 161-173.

Graves, P., Murdoch, J.C., Thayer, M.A. and Waldman, D. (1988), 'The robustness of the hedonic price function - urban air quality', *Land Economics.* Vol 64.3. pp. 220-233.

Grether, D.M. and Miesszkowski, P. (1980), 'The effects of non-residential landuses on the price of adjacent housing: some estimates of proximity effects', *Journal of Urban Economics.* Vol. 8. pp. 1-15.

Griffith, D.A. (1981), 'Evaluating the transformation from a monocentric to a polycentric city', *Professional Geographer.* Vol. 33.2. pp. 189-196.

Griliches, Z. (1971), *Price Indexes and Quality Change,* Harvard University Press.

Gross, D.J. (1988), 'Estimating willingness to pay for housing characteristics: An application of the Ellickson bid-rent model', *Journal of Urban Economics.* Vol 24. pp. 95-112.

Halvorsen, R. and Pollakowski, H. (1982), 'Choice of functional form for hedonic price equations', *Journal of Urban Economics.* Vol. 10. pp. 37-49.

Hamilton, B. (1982), 'Wasteful commuting', *Journal of Political Economy.* Vol. 90. pp. 1035-1053.

Harrison, D.J. and Rubinfeld, D.L. (1978), 'Hedonic housing prices and the demand for clean air', *Journal of Environmental Economics and Management.* Vol 5. pp. 81-102.

Harvey, D. (1972), 'Society, the city and the space-economy of urbanism', *Resource Paper No. 18. Association of American Geography.* Washington D.C.

Harvey, D. (1973), *Social Justice and the City,* Edward Arnold Ltd.

Heikkilla, E. (1988), 'Multicollinearity in regression models with multiple distance measures', *Journal of Regional Science*. Vol. 28. pp. 345-361.

Heikkilla, E., Gordon, P., Kim, J., Deiser, R.B. and Richardson, H.W. (1989), 'What happened to the CBD-distance gradient? Land values in a poliocentric city', *Environment and Planning A*. Vol. 21. pp. 221-232.

Herrin, W. and Kern, C. (1992), 'Testing the standard urban model of residential choice: An implicit markets approach', *Journal of Urban Economics*. Vol. 31. pp. 145-163.

Higgs, G. and Martin, D. (1995), 'A Comparison of Recent Spatial Referencing Approaches in Planning', *Papers in Planning Research*. No. 155, Department of City and Regional Planning. University of Cardiff.

Hirschfield, A., Brown, P. and Todd, P. (1995), 'GIS and the analysis of spatially referenced crime data: experiences in Merseyside, UK', *International Journal of Geographic Information Systems*. Vol 9.2. pp 191-210.

Hughes, W. and Sirmans, C.F. (1992), 'Traffic externalities and single-family house prices', *Journal of Regional Science*. Vol. 32.4. pp. 487-500.

Jackson, J.R. (1979), 'Intraurban variations in the price of housing', *Journal of Urban Economics'*, Vol 6. pp. 464-479.

Johnson, R.C and Kaserman, D.L. (1983), 'Housing market capitalization of energy-saving durable good investments', *Economic Inquiry*. Vol. 21.3 pp. 374-386.

Johnston, R.J., Gregory, D. and Smith, D.M. (1994), *The Dictionary of Human Geography,* Blackwell Publishers Ltd.

Jones, K. (1991), *Multi-level Models for Geographical Research,* Concepts and Techniques in Modern Geography 54, Norwich: Geo Books.

Jones, K. and Bullen, N. (1993), 'A Multilevel analysis of the variations in domestic property prices: Southern England 1980-1987', *Urban Studies*. Vol 30.8. pp. 1409-1426.

Jones, K. and Bullen, N. (1994), 'Contextual models of urban house prices: A comparison of fixed - and random- coefficient models developed by expansion', *Economic Geography*, Vol. 70.3. pp. 252-272.

Kain, J.F. and Quigley, J.M. (1970a), 'Measuring the value of housing quality', *Journal of the American Statistical Association*.

Kain, J.F. and Quigley, J.M. (1970b), 'Evaluating the quality of the residential environment', *Environment and Planning A* Vol. 2. pp. 23-32.

Keltics (1989), *The Cardiff Housing Condition Survey. Phase 1: Inner Area Final Report,* Keltics, Consulting Architects and Engineers Ltd.

Kennedy, P. (1985), *A Guide to Econometrics*, Basil Blackwell. Second Edition.

Kennedy, G.A., Dai, M., Henning, S.A., and Vendeveer, L.R. (1996), 'A GIS-based approach for including topographic and locational attributes in the hedonic analysis of rural land values', *American Journal of Agricultural Economics.* Vol. 78.5. pp. 1419-1437.

Knox, P.L. (1995), *Urban Social Geography: An Introduction,* Routledge. Third edition.

Krumm, R.J. (1980), 'Neighbourhood amenities: an economic analysis', *Journal of Urban Economics.* Vol. 7. pp. 208-224.

Lake, I. (1996), *Using Ordnance Survey digital map data to implement the hedonic pricing method: Assessing visual and road transport disamenity,* University of East Anglia, School of Environmental Studies.

Lancaster, K. (1966), 'A new approach to consumer theory', *Journal of Political Economy.* Vol. 74. pp 132-157.

Lansford, N.H. and Jones, L.L. (1995), 'Recreational and aesthetic value of water using hedonic price analysis', *Journal of Agricultural and Resource Economics.* Vol. 20.2. pp. 341-355.

Lerman, S.R. and Kern, C.R. (1983), 'Hedonic theory, bid-rent, and willingness-to-pay: some extensions of Ellicksons results', *Journal of Urban Economics.* Vol 13. pp. 358-363.

Levesque, T. (1994), 'Modelling the effects of airport noise on residential housing markets', *Journal of Transport Economics and Policy.* Vol 28.2. pp. 199-211.

Li, M.M. and Brown, H.J. (1980), 'Micro-neighbourhood externalities and hedonic house prices', *Land Economics.* Vol. 56. pp. 125-140.

Linnenman, P. (1980), 'Some empirical results on the nature of the hedonic price function for urban housing markets', *Journal of Urban Economics.* Vol 8. pp. 47-68.

Longley, P.A. and Clarke, G. (1995), *GIS for Business and Service Planning,* Pearson Professional Ltd.

Longley, P.A. and Dunn, R. (1988) 'Graphical assessment of housing market models', *Urban Studies,* Vol. 25.1, pp. 21-33.

Longley, P., Higgs, G. and Martin, D. (1994), 'The predictive use of GIS to model property valuations', *International Journal of Geographical Information Systems.* Vol. 8.2. pp. 217-235.

Longley, P., Martin, D. and Higgs, G. (1993), 'The geographical implications of changing local taxation regimes', *Transactions. Institute of British Geographers.* Vol 18. pp. 86-101.

MacDonald, J.F. (1979), *Economic Analysis of an Urban Housing Market Studies in Urban Economics,* Academic Press.

Maclennan, D. (1982), *Housing Economics: An Applied Approach,* Longman: London and New York.

Maddala, G.S. (1992), *Introduction to Econometrics,* Academic Press, Second Edition.

Martin, D. (1992), 'Postcodes and the 1991 Census of population: issues, problems and prospects', *Transactions of the Institute of British Geographers.* Vol 17. pp. 350-357.

Martin, D. (1996), *Geographic Information Systems and Their Socio-Economic Applications,* Routledge, London. Second Edition.

Martin, D. and Higgs, G. (1994), *Georeferencing People and Places: a Comparison of Detailed Data Sets,* Papers in Planning Research. Department of City and Regional Planning. University of Cardiff.

Martin, D., Longley, P. and Higgs, G. (1992), 'The geographical incidence of local government revenues: an intraurban case study', *Environment and Planning C.* Vol 10. pp. 253-265.

Mather, P.M. (1976), *Computational Methods of Multivariate Analysis in Physical Geography,* John Wiley and Sons.

McLeod, P.B. (1984), 'The demand for local amenity: An hedonic price analysis', *Environment and Planning A.* Vol. 16. pp. 359-400.

McMillan, M., Reid, B.G. and Gillen, D.W. (1980), 'An extension of the hedonic approach for estimating the value of quiet', *Land Economics.* Vol 56.3. pp. 315-327.

Merrett, S. and Gary, F. (1982), *'Owner Occupation in Britain',* London: Routledge and Kegan Paul.

Millington, A.F. (1990), *'An Introduction to Property Valuation',* London: Estates Gazette.

Mills, E.S. (1972), *Urban Economics,* Scott, Foresman and Company: London.

Mingche, M.L. and Brown, J.H. (1980), 'Micro-neighbourhood externalities and hedonic housing price', *Land Economics.* Vol 56. pp. 125-141.

Munro, M. and Maclennan, D. (1987), 'Intra-urban changes in housing prices: Glasgow 1972-83', *Housing Studies.* Vol 2.2. pp. 65-81.

Muth, R.F. (1969), *Cities and Housing: The Spatial Pattern of Urban Residential Landuse,* The University of Chicago Press.

Muth, R.F. and Goodman, A.C. (1989), *The Economics of Housing Markets,* Fundamental and Applied Economics 31. Harwood Academic Publishers 1989.

Nelson, J.P (1980), 'Airports and property values', *Journal of Transport Economics and Policy.* Vol 14.1. pp. 37-52.

Nicol, C. (1996), 'Interpretation and compatibility of house price series', *Environment and Planning A.* Vol 28.1. pp. 119-133.

North, J.H. and Griffin, C.C. (1993), 'Water sources as a housing characteristics - hedonic property valuation and willingness-to-pay for water', *Water Resources Research.* Vol. 29.7. pp. 1923-1929.

Odland, J., Golledge, R.G. and Rogerson, P.A (1989), 'Mathematical and statistical analysis in human geography' in G.L. Gailey and J. Willmott (eds), *Geography in America,* Merrill Publishing Company. pp. 719-745.

Ohsfeldt, R.L. (1988), 'Implicit markets and the demand for housing characteristics', *Regional Science and Urban Economics.* Vol. 18. pp. 321-343.

Openshaw, S. (1984) *The Modifiable Areal Unit Problem,* Concepts and Techniques in Modern Geography 38 Norwich: Geo Books.

Openshaw, S (1991) 'Developing appropriate spatial analysis methods for GIS' in D.J. Maguire, M.F. Goodchild and D.W.Rhind (eds), *Geographical Information Systems: Methodology and Potential Applications,* Longmans. London. pp. 389-402.

Openshaw, S. (1994a) 'What is GISable spatial analysis?', *New Tools for Spatial Analysis.* Eurostat, Luxembourg. pp. 36-44.

Openshaw, S. (1994b) 'Two exploratory space-time-attribute pattern analysers relevant to GIS', in Fotheringham, S. and Rogerson, P. (eds), *Spatial Analysis and GIS,* Taylor and Francis Ltd. pp. 83-104.

Openshaw, S. (1995) 'Developing GIS relevant zone based spatial analysis methods', *Paper at the Institute British Geographers Annual Conference.* University of Newcastle.

Openshaw, S. and Taylor. P.J. (1979), 'A million or so correlation coefficients: three experiments on the modifiable areal unit problem', in R.J. Bennett, N.J. Thrift, and N. Wrigley (eds), *Statistical Applications in the Spatial Sciences,* London: Pion.

Ordnance Survey (1993) *ADDRESS-POINT User Guide*, Ordnance Survey, Southampton.

Orford, S., Harris, R. and Dorling, D. (1999), 'Information Visualization in the Social Sciences', *Social Science Computer Review.* 17. 2.

Orford, S. (1998), 'Valuing Location in an Urban Housing Market', *Proceedings of the 3rd International Conference on Gecomputation, Bristol, UK, 17-19 September 1998*, GeoComputation CD-ROM. ISBN 0-9533477-0-2.

Ozanne, L. and Malpezzi, S. (1985), 'The efficacy of hedonic model estimation with the annual housing survey - evidence from demand experiments' *Journal of Economic and Social Measurement.* Vol. 13.2. pp. 153-172.

Palmquist, R.B. (1984), 'Estimating the demand for the characteristics of housing', *Review of Economics and Statistics*. No. 66. pp. 394-404.

Papageorgiou, G.J. (1976), *Mathematical Landuse Theory,* Lexington Books.

Pinch, S. (1985), *'Cities and services: The Geography of Collective Consumption',* Routledge and Kegan Paul.

Pindyck, R.S. and Rubinfeld, D. (1991), *'Econometric Models and Economic Forecasts',* McGraw-Hill International Editions. Economic Series. 3rd Edition.

Powe, N.A., Garrod, D., and Willis, K.G. (1995), 'Valuation of urban amenities using an hedonic price model', *Journal of Property Research*. Vol. 12. pp. 137-147.

Quigley, J.M. (1982), 'Non-linear budget constraints and consumer demand - an application to public programs for residential housing', *Journal of Urban Economics*. Vol. 12.2. pp. 177-201.

Raper, J., Rhind, D. and Shepard, J. (1992), *Postcodes: the New Geography,* Longman, London.

Rasbash, J. and Woodhouse, G. (1995), *Mln Command Reference,* London, Multilevel Models Project, Institute of Education.

Richardson, H.W. (1976), 'Relevance of mathematical land use theory to application', in George J. Papageorgiou (ed.), *Mathematical Land Use Theory,* Lexington Books.

Ridker, R. and Henning, J. (1968), 'The determination of residential property values with special reference to air pollution', *Review of Economics and Statistics*.

Rogerson, P.A. and Fotheringham, A.S. (1994), 'GIS and spatial analysis: Introduction and overview', in Fotheringham, S. and Rogerson, P. (eds), *Spatial Analysis and GIS,* Taylor and Francis Ltd. pp. 1-10.

Rosen, S. (1974), 'Hedonic prices and implicit markets: Product differentiation in pure competitions', *Journal of Political Economy*. Vol. 72. pp. 34-55.

Sanchez, T.W. (1993), 'Measuring the impact of highways and roads on residential property values with GIS: An empirical analysis of Atlanta', in Klosterman, R.E and French, S.P (eds), *Third International Conference on Computers in Urban Planning and Management*. Vol. 2.

Saunders, P. (1990), *A Nation of Home Owners*, Unwin Hyman Ltd.

Schnare, A. and Struyk, R.J. (1976), 'Segmentation in urban housing markets', *Journal of Urban Economics*. Vol. 3. pp. 146-166.

Schroeder, L.D., Sjoquist, D.L. and Stephan, P.E. (1986), *Understanding Regression Analysis: An Introductory Guide,* Sage Publications.

Short, J.R. (1982), *Housing in Britain: The Post-War Experience*, Methuen.

Sinton, D.F (1992), 'Reflections on twenty five years of GIS', Insert in *GIS World*. February 1992. pp. 1-8.

Smith, B. (1996), 'The NLIS in 1996: the pilot project expands', *Mapping Awareness*. Vol. 10.2. pp. 22-24.

Straszheim, M. (1974), 'Hedonic estimation of housing market prices. A further comment', *Review of Economics and Statistics*. pp. 404-406.

Stull, W.J. (1975), 'Community environment, zoning, and the market value of single family homes', *Journal of Law Economics*. Vol. 18. pp. 535-558.

Tyrvainen, L. (1997), 'The amenity value of the urban forest: An application of the hedonic pricing method', *Landscape and Urban Planning*. Vol. 37.3. pp. 211-222.

Veldhuisen, K.J. and Timmermans, H.P. (1984), 'Specification of individual residential utility-functions - a comparative analysis of three measurement procedures', *Environment and Planning A*. Vol. 16.12. pp. 1573-1582.

Wabe, J.S. (1971), 'A study of house prices as a means of estimating the value of journey time, the rate of time preferences and the valuation of some aspects of environment in the London Metropolitan Region', *Applied Economics*. Vol. 3. pp. 247-255.

Waddell, P. and Berry, B.J.L (1993), 'Housing price gradients - the intersection of space and built form', *Geographical Analysis*. Vol. 25.1. pp. 5-19.

Waddell, P., Berry, B.J.L. and Hoch, I. (1993), 'Residential property values in a multinodal urban area: New evidence on the implicit price of location', *Journal of Real Estate Finance and Economics*. Vol 7. pp. 117-141.

Wilkinson, R.K. (1971), 'The determinants of relative house prices', *Urban Studies*. Vol. 8, pp. 50-59.

Wilkinson, R.K. (1973a), 'House prices and the measurement of externalities', *Economic Journal*. Vol 83, pp. 72-86.

Wilkinson, R.K. (1973b), 'Measuring the determinants of relative house prices', *Environment and Planning*. Vol 5, pp. 357-367.

Wilkinson, R.K. (1974), 'The determinants of relative house prices: a case of academic astigmatism', *Urban Studies*. Vol. 11, pp. 227-230.

Witte, A. Sumka, J. and Erckson, H. (1979), 'An estimation of a structural hedonic price model of the housing market: An application of Rosen's theory of implicit markets', *Econometrica*. Vol. 47. pp. 1151-1173.

Woodhouse, G., Rasbash, J., Goldstein, H. and Yang, M. (1996), 'Introduction to multi-level modelling' in Woodhouse, G. (eds) *Multi-*

Level Modelling Applications: A Guide for Users of Mln, Institute of Education, University of London.

Wrigley, N. (1995), 'Revisiting the MAUP and the ecological fallacy', in Cliff, A.D., Hoare, A.G. and Thrift, N.J. (eds), *Diffusing Geography,* Blackwell.

Wyatt, P. (1994), 'Using a GIS for property valuation', *Association of American Geographers,* pp. 11-13.

Index